# A Smarter National Surveillance System for Occupational Safety and Health in the 21st Century

Committee on Developing a Smarter National Surveillance System for Occupational Safety and Health in the 21st Century

Board on Agriculture and Natural Resources
Division on Earth and Life Studies

Committee on National Statistics
Division of Behavioral and Social Sciences and Education

Board on Health Sciences Policy
Health and Medicine Division

A Consensus Study Report of

*The National Academies of*
SCIENCES · ENGINEERING · MEDICINE

THE NATIONAL ACADEMIES PRESS
*Washington, DC*
**www.nap.edu**

**THE NATIONAL ACADEMIES PRESS 500 Fifth Street, NW Washington, DC 20001**

This activity was supported by Grant 200-2011-38807 from the Centers for Disease Control and Prevention, National Institute for Occupational Safety and Health, Grant HHSP233201400020B from the Department of Labor, Bureau of Labor Statistics, and Grant DOL-OPS-16-P-000193 from Department of Labor, Occupational Safety and Health Administration. Any opinions, findings, conclusions, or recommendations expressed in this publication do not necessarily reflect the views of any organization or agency that provided support for the project.

International Standard Book Number-13: 978-0-309-46299-0
International Standard Book Number-10: 0-309-46299-1
Digital Object Identifier: https://doi.org/10.17226/24835

Additional copies of this report are available for sale from the National Academies Press, 500 Fifth Street NW, Keck 360, Washington, DC 20001; (800) 624-6242 or (202) 334-3313; http://www.nap.edu.



Printed in the United States of America

Suggested citation: National Academies of Sciences, Engineering, and Medicine. 2018. *A Smarter National Surveillance System for Occupational Safety and Health in the 21st Century*. Washington, DC: The National Academies Press. doi: https://doi.org/10.17226/24835.

*The National Academies of*

# SCIENCES • ENGINEERING • MEDICINE

The **National Academy of Sciences** was established in 1863 by an Act of Congress, signed by President Lincoln, as a private, nongovernmental institution to advise the nation on issues related to science and technology. Members are elected by their peers for outstanding contributions to research. Dr. Marcia McNutt is president.

The **National Academy of Engineering** was established in 1964 under the charter of the National Academy of Sciences to bring the practices of engineering to advising the nation. Members are elected by their peers for extraordinary contributions to engineering. Dr. C. D. Mote, Jr., is president.

The **National Academy of Medicine** (formerly the Institute of Medicine) was established in 1970 under the charter of the National Academy of Sciences to advise the nation on medical and health issues. Members are elected by their peers for distinguished contributions to medicine and health. Dr. Victor J. Dzau is president.

The three Academies work together as the **National Academies of Sciences, Engineering, and Medicine** to provide independent, objective analysis and advice to the nation and conduct other activities to solve complex problems and inform public policy decisions. The National Academies also encourage education and research, recognize outstanding contributions to knowledge, and increase public understanding in matters of science, engineering, and medicine.

Learn more about the National Academies of Sciences, Engineering, and Medicine at **www.nationalacademies.org**.

*The National Academies of*

SCIENCES • ENGINEERING • MEDICINE

**Consensus Study Reports** published by the National Academies of Sciences, Engineering, and Medicine document the evidence-based consensus on the study's statement of task by an authoring committee of experts. Reports typically include findings, conclusions, and recommendations based on information gathered by the committee and the committee's deliberations. Each report has been subjected to a rigorous and independent peer-review process and it represents the position of the National Academies on the statement of task.

**Proceedings** published by the National Academies of Sciences, Engineering, and Medicine chronicle the presentations and discussions at a workshop, symposium, or other event convened by the National Academies. The statements and opinions contained in proceedings are those of the participants and are not endorsed by other participants, the planning committee, or the National Academies.

For information about other products and activities of the National Academies, please visit www.nationalacademies.org/about/whatwedo.

## COMMITTEE ON DEVELOPING A SMARTER NATIONAL SURVEILLANCE SYSTEM FOR OCCUPATIONAL SAFETY AND HEALTH IN THE 21ST CENTURY

*Chair*

EDWARD H. SHORTLIFFE, NAM,[1] Arizona State University, Tempe

*Members*

DAVID K. BONAUTO, Washington State Department of Labor and Industries, Tumwater
DAVID L. BUCKERIDGE, McGill University, Montreal, Quebec, Canada
STEVEN B. COHEN, RTI International, Washington, DC
LETITIA K. DAVIS, Massachusetts Department of Public Health, Boston
GERALD F. KOMINSKI, University of California, Los Angeles
SCOTT A. MUGNO, FedEx Ground, Moon Township, PA
KENNETH D. ROSENMAN, Michigan State University, East Lansing
NOAH S. SEIXAS, University of Washington School of Public Health, Seattle
MARGARET M. SEMINARIO, AFL-CIO, Washington, DC
GLENN M. SHOR, California Department of Industrial Relations, Oakland
DAVID H. WEGMAN, University of Massachusetts Lowell

*Staff*

PEGGY TSAI YIH, Study Director, Board on Agriculture and Natural Resources
JENNA BRISCOE, Research Assistant, Board on Agriculture and Natural Resources
CATHY LIVERMAN, Scholar, Board on Health Sciences Policy
MICHAEL COHEN, Senior Program Officer, Committee on National Statistics
YASMIN ROMITTI, Research Associate, Board on Atmospheric Sciences and Climate, Board on Earth Sciences and Resources
ROBIN SCHOEN, Director, Board on Agriculture and Natural Resources

---

[1]National Academy of Medicine.

## BOARD ON AGRICULTURE AND NATURAL RESOURCES

*Chair*

CHARLES W. RICE, Kansas State University, Manhattan

*Members*

SHANE C. BURGESS, University of Arizona, Tucson
SUSAN CAPALBO, Oregon State University, Corvallis
GAIL CZARNECKI-MAULDEN, Nestle Purina PetCare, St. Louis, MO
GEBISA EJETA, Purdue University, West Lafayette, IN
JAMES S. FAMIGLIETTI, California Institute of Technology, Pasadena
FRED GOULD, NAS,[1] North Carolina State University, Raleigh
DOUGLAS B. JACKSON-SMITH, Ohio State University, Wooster
MOLLY M. JAHN, University of Wisconsin–Madison
ROBBIN S. JOHNSON, Cargill Foundation, Wayzata, MN
JAMES W. JONES, NAE,[2] University of Florida, Gainesville
A.G. KAWAMURA, Solutions from the Land, Washington, DC
STEPHEN S. KELLEY, North Carolina State University, Raleigh
JULIA L. KORNEGAY, North Carolina State University, Raleigh
JAN E. LEACH, Colorado State University, Fort Collins
JILL J. MCCLUSKEY, Washington State University, Richland
KAREN I. PLAUT, Purdue University, West Lafayette, IN
JIM E. RIVIERE, NAM,[3] Kansas State University, Manhattan

*Staff*

ROBIN A. SCHOEN, Director

[1]National Academy of Sciences.
[2]National Academy of Engineering.
[3]National Academy of Medicine.

## COMMITTEE ON NATIONAL STATISTICS

*Chair*

ROBERT M. GROVES, NAS,[1] NAM,[2] Georgetown University, Washington, DC

*Members*

FRANCINE BLAU, Cornell University, Ithaca, NY
MARY ELLEN BOCK, Purdue University, West Lafayette, IN
ANNE E. CASE, NAM,[2] Princeton University, Princeton, NJ
MICHAEL CHERNEW, NAM,[2] Harvard Medical School, Boston, MA
JANET CURRIE, NAM,[2] Princeton University, Princeton, NJ
DONALD A. DILLMAN, Washington State University, Pullman
CONSTANTINE GATSONIS, Brown University, Providence, RI
JAMES S. HOUSE, NAM,[2] University of Michigan, Ann Arbor
THOMAS L. MESENBOURG, Retired; formerly, U.S. Census Bureau
SARAH M. NUSSER, Iowa State University, Ames
COLM A. O'MUIRCHEARTAIGH, University of Chicago, IL
JEROME P. REITER, Duke University, Durham, NC
ROBERTO RIGOBON, Massachusetts Institute of Technology, Cambridge
JUDITH A. SELTZER, University of California, Los Angeles
EDWARD H. SHORTLIFFE, NAM,[2] Arizona State University, Tempe

*Staff*

BRIAN A. HARRIS-KOJETIN, Director

[1]National Academy of Sciences.
[2]National Academy of Medicine.

## BOARD ON HEALTH SCIENCES POLICY

*Chair*

Jeffrey Kahn, Johns Hopkins University, Baltimore, MD

*Members*

Wylie Burke, NAM,[1] University of Washington, Seattle
R. A. Charo, NAM,[1] University of Wisconsin–Madison
Linda H. Clever, NAM,[1] California Pacific Medical Center, San Francisco
Barry S. Coller, NAS,[2] NAM,[1] The Rockefeller University, New York, NY
Bernard A. Harris, Jr., Vesalius Ventures, Houston, TX
Martha N. Hill, NAM,[1] Johns Hopkins University School of Nursing, Baltimore, MD
Steven E. Hyman, NAM,[1] Broad Institute, Cambridge, MA
Alan M. Jette, NAM,[1] Boston University School of Public Health, MA
Patricia A. King, NAM,[1] Georgetown University Law Center, Washington, DC
Story C. Landis, NAM,[1] National Institute of Neurological Disorders and Stroke, Bethesda, MD
Harry T. Orr, NAM,[1] University of Minnesota, Minneapolis
Umair A. Shah, Harris County Public Health and Environmental Services, Houston, TX
Robyn Stone, NAM,[1] LeadingAge, Washington, DC
Sharon Terry, Genetic Alliance, Washington, DC

*Staff*

Andrew Pope, Director

[1]National Academy of Medicine.
[2]National Academy of Sciences.

# Preface

Many threats to health and well-being occur in the workplace. Understanding those risks to prevent injury, illness, or even fatal incidents is an important function of society. We expect interested parties to measure the frequency of incidents, to determine causes when possible, and to offer preventive measures to stakeholders so that the work environment becomes safer and healthier over time. This all needs to occur in the context of a changing workforce and the evolution of the nature of work, suggesting that new kinds of threats to health and well-being can occur, even as others are being optimally managed or are becoming less pertinent as jobs or industries fade away and are replaced by others. In the United States, both the federal and state governments are heavily involved in monitoring occupational health and safety and developing policies or interventions intended to mitigate work-related injuries and disease. At the federal level, principal agencies involved with such work include the Bureau of Labor Statistics (BLS), the National Institute for Occupational Safety and Health (NIOSH, a division of the Centers for Disease Control and Prevention), and the Occupational Safety and Health Administration (OSHA).

Many of the challenges faced by these agencies relate to gathering the information necessary to measure and assess the frequency of workplace-related injuries and illnesses so that suitable policies and interventions can be proposed. In 1987, the agencies sought the advice of the National Research Council regarding surveillance for occupational safety and health (OSH). The resulting report, *Counting Injuries and Illnesses in the Workplace: Proposals for a Better System*, provided seminal guidance on how to organize and enhance the U.S. programs for OSH surveillance. In the subsequent 30 years, despite many efforts developed in response to the 1987 recommendations, both the technology for surveillance and the nature of work have evolved considerably.

In the intervening years, the National Academies of Sciences, Engineering, and Medicine have been involved in many workshops, studies, and reports that have touched on issues related to occupational safety and health. For example, from 2006-2009, NIOSH requested that the National Academies undertake programmatic reviews of some NIOSH research programs. The need for better surveillance was a theme that emerged from each of those reviews. In 2014, sequestration forced federal agencies to re-examine their programs as they were asked to address their agency's mission more effectively but with fewer resources. More recently, the director of NIOSH again approached the National Academies to assist them and other agencies to come up with creative solutions across many categories of occupation for surveillance and monitoring, and to be able to do so in a "smarter" and cost-effective manner.

## THE COMMITTEE AND ITS CHARGE

To obtain forward-looking advice, NIOSH, BLS, and OSHA jointly asked the National Academies to conduct a study in response to the need for a more coordinated, cost-effective set of approaches for occupational safety and health surveillance in the United States. Our study committee has addressed this task, gathering information about the strengths and limitations of existing national and state approaches, reviewing a variety of methodologies and technologies that might be applied usefully and cost-effectively. The resulting report is a product of more than a year of deliberations, offering the consensus advice of a diverse set of individuals who have studied the issues carefully and learned a great deal in the process. We have formulated a future vision that is intended to assist all stakeholders, including the agencies, as they seek to improve occupational safety and health in the coming years.

Some of the committee members are career professionals who have worked in the area of occupational safety and health, both at state and national levels. Others brought complementary skill sets that were pertinent to the committee's charge: epidemiology, occupational medicine, survey methodologies, biomedical informatics, data analytics, economics, cost-benefit analysis, and workplace organization and management. I was honored to help lead this diverse group of talented professionals, all of whom contributed enthusiastically and tirelessly to the discussions, deliberations, and the final report. We quickly realized that the topic is very large and complex, with nuances that many of us had not anticipated when we joined the study group. Our knowledge of the topic was broadened, in the first three meetings and several conference calls during the early months, by informative sessions with invited experts who helped us to address the task. Large portions of the early meetings were open, with members of the public invited to attend and to provide comments.

We turned, for our final two meetings, to private deliberations, reviewing all that we had learned in order to develop a shared vision of what was possible and seeking to offer recommendations that were responsive to the committee's charge and were actionable. Our task was further influenced by a change in government during the study period, leading to uncertainties about future budgets and focus for the pertinent federal agencies.

## ACKNOWLEDGMENTS

This report benefited from the combined talents of many people, including those who were directly associated with the project and many who were not. First, thanks are due to members of the committee itself, all of whom maintained a high level of enthusiasm, energy, and dedication over the course of the project. Committee members found time for five project meetings, for multiple conference calls, and for drafting portions of the text despite their many other responsibilities and commitments.

Many other people volunteered their time and expertise to help the committee to understand better the ways in which OSH surveillance might be strengthened and modernized in ways that would be acceptable, and even inspirational, for all stakeholders. The information gathered during these interactions proved invaluable to the committee's deliberations and forms the backbone of this report. Those who provided us with briefings are summarized in Appendix C to the report. We thank them all for their important contributions.

The committee members, many of whom have served on previous study groups, stand in awe of the remarkable, patient work of the National Academies staff in supporting its deliberations over the course of this study. Staff members kept the committee on track and helped its members to put their ideas and analyses into coherent prose.

Our thanks, finally, to representatives of the sponsoring agencies who worked closely with us when we needed information. Not only did they offer insights in the public sessions during the first committee meetings, but they responded fully and promptly to information requests that we submitted to them throughout the study. The committee accordingly hopes that its findings and recommendations will assist the agencies in moving forward to assure the design and implementation of a "smarter" OSH surveillance system for the 21st century.

Edward H. Shortliffe, *Chair*
Committee on Developing a "Smarter" National Surveillance System for Occupational Safety and Health in the 21st Century

# Acknowledgments

This Consensus Study Report was reviewed in draft form by individuals chosen for their diverse perspectives and technical expertise. The purpose of this independent review is to provide candid and critical comments that will assist the National Academies of Sciences, Engineering, and Medicine in making each published report as sound as possible and to ensure that it meets the institutional standards for quality, objectivity, evidence, and responsiveness to the study charge. The review comments and draft manuscript remain confidential to protect the integrity of the deliberative process.

We thank the following individuals for their review of this report:

James J. Cimino, University of Alabama School of Medicine
Eric Frumin, Change to Win
Erica Groshen, Cornell University
James S. House, University of Michigan
Joel Kaufman, University of Washington
Anthony LaMontagne, Deakin University, Australia
Harold Lehmann, Johns Hopkins School of Medicine
Virginia Lesser, Oregon State Department of Statistics
Sharon Levine, Kaiser Permanente
Barbara Materna, California Department of Public Health
Carrie A. Redlich, Yale University
John W. Ruser, Workers Compensation Research Institute
Gregory R. Wagner, Harvard School of Public Health
Marc Younis, ExxonMobil

Although the reviewers listed above provided many constructive comments and suggestions, they were not asked to endorse the conclusions or recommendations of this report nor did they see the final draft before its release. The review of this report was overseen by **Dr. Mark R. Cullen, Stanford University,** and **Dr. James A. Merchant, the University of Iowa.** They were responsible for making certain that an independent examination of this report was carried out in accordance with the standards of the National Academies and that all review comments were carefully considered. Responsibility for the final content rests entirely with the authoring committee and the National Academies.

# Contents

# Select Acronyms and Abbreviations

| | |
|---|---|
| ABLES | Adult Blood Lead Epidemiology and Surveillance |
| ACS | American Community Survey |
| AHRQ | Agency for Healthcare Research and Quality |
| ATS | American Thoracic Society |
| | |
| BLL | blood lead level |
| BLS | Bureau of Labor Statistics, U.S. Department of Labor |
| BRFSS | Behavioral Risk Factor Surveillanced System |
| | |
| CalSIM | California Simulation of Insurance Markets |
| CAREX | CARcinogen EXposure System |
| CDC | Centers for Disease Control and Prevention |
| CEA | Cost-effectiveness Analysis |
| CEHD | Chemical Exposure Health Data |
| CFID | Commercial Fishing Incident Database |
| CFOI | Census of Fatal Occupational Injuries |
| CFR | Code of Federal Regulations |
| CIPSEA | Confidential Information Protection and Statistical Efficiency Act |
| COPD | Chronic Obstructive Pulmonary Disease |
| CPSC | Consumer Products Safety Commission |
| CPWR | The Center for Construction Research and Training |
| CSELS | Center for Surveillance, Epidemiology, and Laboratory Services |
| CSTE | Council of State and Territorial Epidemiologists |

| | |
|---|---|
| CUA | Cost-utility Analysis |
| CWCS | Center for Workers' Compensation Studies |
| CWHSP | Coal Workers' Health Surveillance Program |
| CWP | Coal Workers' Pneumoconiosis |
| CWXSP | Coal Workers' X-ray Surveillance Program |
| | |
| DAFW | days away from work |
| DOE | U.S. Department of Energy |
| DOL | U.S. Department of Labor |
| DSR | Division of Safety Research, National Institute for Occupational Safety and Health |
| | |
| ED | emergency department |
| EHR | electronic health record |
| EPA | Environmental Protection Agency |
| EPIDERM | skin specialist surveillance scheme |
| EU | European Union |
| Eurofound | European Foundation for the Improvement of Living and Working Conditions |
| EWCS | European Working Conditions Survey |
| eWorld | Work-Related Lung Disease Surveillance System |
| | |
| FAA | Federal Aviation Administration |
| FACE | Fatality Assessment and Control Evaluation |
| FARS | Fatality Analysis Reporting System |
| FINJEM | Finnish Job Exposure Matrix |
| FOG | Fatalities in the Oil and Gas Extraction Industry |
| FRA | Federal Railroad Administration |
| FTE | full-time equivalent |
| | |
| GP | general practitioner |
| | |
| HCUP | The Healthcare Cost and Utilization Project |
| HHE | health hazard evaluation |
| HHS | U.S. Department of Health and Human Services |
| HITECH | Health Information Technology for Economic and Clinical Health |
| HIV | human immunodeficiency virus |
| HSE | Health and Safety Executive |
| HSOII | Household Survey of Occupational Injuries and Illnesses |
| | |
| IAIABC | International Association of Industrial Accidents Boards and Commissions |

| | |
|---|---|
| ICD | International Classification of Diseases |
| ICD-10-CM | International Classification of Diseases, 10th Revision, Clinical Modification |
| ILO | International Labour Organization |
| IMIS | Integrated Management Information System |
| IOM | Institute of Medicine |
| IT | information technology |
| | |
| JEMs | job exposure matrices |
| | |
| LAUS | Local Area Unemployment Statistics |
| LFS | Labour Force Survey |
| | |
| MA COSH | Massachusetts Coalition for Occupational Safety and Health |
| MC | methylene chloride |
| MDRS | Mine Data Retrieval System |
| MEPS | Medical Expenditure Panel Survey |
| MIFACE | Michigan Fatality Assessment and Control Evaluation |
| MSD | musculoskeletal disorder |
| MSHA | Mine Safety and Health Administration |
| | |
| NAICS | North American Industry Classification System |
| NBDPS | National Birth Defects Prevention Study |
| NCCDP | National Center for Chronic Disease Prevention and Health Promotion |
| NCCI | National Council on Compensation Insurance |
| NCHS | National Center for Health Statistics |
| NCIPC | National Center for Injury Prevention and Control |
| NCS | National Compensation Survey |
| NCVS | National Crime Victimization Survey |
| NEISS | National Electronic Injury Surveillance System |
| NHANES | National Health and Nutrition Examination Survey |
| NHIS | National Health Interview Survey |
| NHTSA | National Highway Traffic Safety Administration |
| NIH | National Institutes of Health |
| NIOCCS | NIOSH Industry and Occupation Computerized Coding System |
| NIOSH | National Institute for Occupational Safety and Health |
| NLP | natural language processing |
| NNDSS | National Notifiable Diseases Surveillance System |
| NOES | National Occupational Exposures Survey |
| NOHS | National Occupational Hazards Survey |

| | |
|---|---|
| NOMS | National Occupational Mortality Surveillance System |
| NORA | National Occupational Research Agenda |
| NORC | National Opinion Research Center |
| NORMS | National Occupational Respiratory Mortality System |
| NRC | National Research Council |
| | |
| O*NET | Occupational Network Database |
| ODI | OSHA Data Initiative |
| OHSN | Occupational Health Safety Network |
| OIICS | Occupational Injury and Illness Classification System |
| OIS | OSHA Information System |
| ONC | Office of the National Coordinator for Health Information Technology |
| ORS | Occupational Requirements Survey |
| OSH | Occupational Safety and Health |
| OSH Act | Occupational Safety and Health Act of 1970 |
| OSHA | Occupational Safety and Health Administration |
| | |
| PPE | personal protective equipment |
| PRAMS | Pregnancy Risk Assessment Monitoring System |
| | |
| QA | Quality Assurance |
| QALY | quality-adjusted life year |
| QCEW | Quarterly Census of Employment and Wages |
| QWS | Quality of Worklife Survey |
| | |
| RIDDOR | Reporting of Injuries, Diseases and Dangerous Occurrences Regulations |
| | |
| SENSOR | Sentinel Event Notification System for Occupational Risks |
| SHARP | Safety and Health Assessment and Research for Prevention |
| SOC | Standard Occupational Classification |
| SOII | Survey of Occupational Injuries and Illnesses |
| SROI | social return on investment |
| SSA | Social Security Administration |
| SSDI | Social Security Disability Insurance |
| SWI | Self-reported Work-related Illness |
| | |
| THOR | The Health and Occupation Research network |
| TIRES | Trucking Injury Reduction Emphasis through Surveillance |
| | |
| UAW | United Automobile, Aerospace and Agricultural Implement Workers of America |

| | |
|---|---|
| UI | Unemployment Insurance |
| UK | United Kingdom |
| UMLS | Unified Medical Language System |
| | |
| WCIO | Workers Compensation Insurance Organizations |
| WISQARS | Web-based Injury Statistics Query and Reporting System |
| WMSD | work-related musculoskeletal disorder |
| Work-RISQS | Work-Related Injury Statistics Query System |
| | |
| YWP | Young Workers Project |

# Summary

## INTRODUCTION

The workplace is where 156 million working adults in the United States spend many waking hours, and it has a profound influence on health and well-being. Work-related injuries and illnesses can impact quality of life. Although some occupations and work-related activities are more hazardous than others and face higher rates of injuries, illness, disease, and fatalities, workers in all occupations face some form of work-related safety and health concerns. Not only are such hazards a threat to worker well-being and hence to the nation's health, but one study estimated the annual cost of occupational injuries, illnesses, and deaths in the United States at $250 billion (in 2007 dollars), which is more than the $219 billion for cancer and more than half of the $431.8 billion for cardiovascular disease. Reducing that health burden is the goal of occupational safety and health (OSH) surveillance.

OSH surveillance provides the data and analyses needed to understand the relationships between work and injuries and illnesses in order to improve worker safety and health and prevent work-related injuries and illnesses. Information about the circumstances in which workers are injured or made ill on the job and how these patterns change over time is essential to develop effective prevention programs and target future research. The nation needs a robust OSH surveillance system to provide this critical information for informing policy development, guiding educational and regulatory activities, developing safer technologies, and enabling research and prevention strategies that serves and protects all workers.

The 1987 National Research Council report *Counting Injuries and Illnesses in the Workplace: Proposals for a Better System* provided initial guidance to organize and enhance OSH surveillance in the United States. Responses to that report resulted in a number of improvements, primarily ones that addressed injury surveillance. While the 1987 report was instrumental to many noted improvements in OSH surveillance, major changes have occurred in the past 30 years that necessitate this new study—a comprehensive reassessment of the state of OSH surveillance. The three key federal agencies involved with occupational safety and health—the National Institute for Occupational Safety and Health (NIOSH), the Bureau of Labor and Statistics (BLS), and the Occupational Safety and Health Administration (OSHA)—called on the National Academies of Sciences, Engineering, and Medicine to undertake a study to develop a vision and steps toward a national surveillance system for occupational safety and health for the 21st century (see Statement of Task in Box 1-1).

## BACKGROUND

Over the past 30 years, there have been remarkable changes in the landscape of work. These include major changes in the geographic and proportional distribution of industries, the nature of work, the demographics of the workforce, and employee-employment arrangements. Employment in manufacturing has declined, while there has been significant growth in employment in the service sector, including health care. Individuals are likely to be working more than one job over their working life, and may hold multiple jobs at the same time. The workforce is much more diverse, with many more women, racial and ethnic minorities, and immigrants employed. Growth has occurred in nonstandard work arrangements (such as the use of independent contractors and the outsourcing of functions to other entities) and in "on-demand" or "gig" work (where employment is characterized by short-term contracts or freelance work). With these shifts in work and the workforce, employment is more precarious and many workers lack the protections and rights afforded by laws and regulations that make them more vulnerable and subject to abuse. Approaches to OSH surveillance have generally not evolved to address the changing nature of work.

OSH surveillance is a collaborative effort of federal, state, and local agencies and stakeholders across employers, employee organizations, professional associations, and other organizations. The federal agencies that play the major roles are BLS, OSHA, and the Mine Safety and Health Administration (all in the Department of Labor), and NIOSH (in the Centers for Disease Control and Prevention (CDC) of the Department of Health and Human Services (HHS)). Since the 1890s, BLS has collected statistical data on work-related injuries, illnesses, and fatalities. The OSH Act of

1970 created both OSHA and NIOSH, charging OSHA with responsibility for setting and enforcing safe and healthful workplace standards and tasking NIOSH with conducting research, experiments, and demonstrations relating to occupational safety and health and with developing criteria for recommended standards. There are a number of other federal agencies with responsibilities and programs pertaining to OSH surveillance and prevention. State agencies also play a critical and complementary role in partnership with federal agencies. State agencies collect, analyze, and disseminate data from local sources to guide preventive action at the state, regional, and local levels; provide aggregated data to federal agencies for national surveillance; and fill in gaps in national surveillance data. The strong role of workers and employers is crucial for ensuring accurate and complete data and for using this information to implement workplace improvements. In addition, health care facilities and organizations, workers' compensation systems, and insurance companies have data that are relevant to OSH.

Surveillance is defined as ongoing, systematic collection, analysis, and interpretation of health data, essential to the planning, implementation, and evaluation of public health practice, closely integrated with the dissemination of these data to those who need to know. There is no single, comprehensive OSH surveillance system in the United States, but rather an evolving set of systems using a variety of data sources that meet different surveillance objectives, each with strengths and weaknesses. The major focus to date has been on collecting data on health outcomes, with less emphasis on collection of data on hazards and exposures.

OSH surveillance activity is also spread across multiple federal and state agencies, as agencies collect, store, analyze, and interpret data to meet their specific agency needs. Resources are limited for OSH surveillance, partly because responsibility for occupational issues is bifurcated between labor and public health and partly because of limited core funding allocations specifically for OSH surveillance. There are also barriers to sharing information among federal and state agencies due to long-established conventions, and also barriers to sharing information between employers and employees, largely due to lack of trust.

A major change has been in the evolution, effectiveness, and relative ubiquity of information and communications technology (both methods and tools) since the 1987 report. Advances in data collection and storage, analytic methods, sensors, and mobile devices allow information to be collected and connected with central collection resources. Social media have also become sources of insight regarding societal trends and offer one of many methods for information dissemination. OSH surveillance efforts will need to leverage newer technologies and tools for identifying, organizing, analyzing, and interpreting data in more innovative, powerful, and cost-effective ways. Doing so could reveal problems, trends, and emerging issues

**BOX S-1**
**Guiding Principles and Objectives of an Ideal National Occupational Safety and Health (OSH) Surveillance System**

*Guiding Principles*

1. Robust and collaborative federal leadership built on strong ties with states and other relevant stakeholders is critical to successful OSH surveillance. Engagement of the community of users who need occupational safety and health surveillance information for action is essential.
2. Continuous monitoring of data quality and program activities is essential to ensure program efficiency and impact.
3. Privacy, confidentiality, and access to data needs to be safeguarded while maximizing the utility of surveillance information for prevention.
4. Timely analysis and interpretation of surveillance inputs with routine dissemination of information in relevant formats promotes the use of surveillance information for action by all stakeholders.
5 An efficient, reliable national surveillance system requires public health professionals with training in occupational injury, illness, and hazard surveillance, with the tools and technology necessary to achieve surveillance objectives.
6. The consistent use of standards for data collection, analysis, and information presentation and dissemination will heighten the efficiency and effectiveness of OSH surveillance.

*Objectives*

1. Guide immediate action to control threats to occupational health and safety.
2. Measure the burden of work-related injuries or illnesses and monitor trends over time and space.
3. Identify industries, occupations, and worksites as well as populations, defined by sociodemographic characteristics or work arrangements at high risk for work-related injury, illness, or hazardous exposures.
4. Detect and respond to new or emerging workplace hazards or facilitate the investigation of new diseases linked to occupational exposures.
5. Guide the planning, implementation, and evaluation of programs and policies intended to prevent and control work-related injuries, illnesses, and hazardous workplace exposures.
6. Generate hypotheses and to make data available for research.

within and across sectors, groups, and geographic regions of workers. Also, these technologies offer opportunities to improve the dissemination of information to those that can use surveillance data to take preventive action, thereby improving worker safety and health and reducing associated human and economic costs of work-related injuries and illnesses.

The *Committee on Developing a "Smarter" National Surveillance System for Occupational Safety and Health in the 21st Century* undertook its task by first considering the goals of an ideal national surveillance system and establishing a set of guiding principles (see Box S-1). Then it examined the current roles and activities of different agencies and stakeholders, and studied OSH surveillance in other countries for possible lessons learned. Next, the committee explored promising new developments, such as the household survey, electronic health records, autocoding of occupational information, electronic reporting, use of workers' compensation data, and improvements in occupational hazard and exposure surveillance. Then the committee considered ways to enable an effective national OSH surveillance system, including a clear rationale and prioritization for surveillance, coordination of surveillance strategies, effective use of information technology, and utilizing practitioners with appropriate skills. Finally, to address the demands and concerns of the current and future workforce, and ultimately to protect workers, the committee envisioned how to achieve a more cohesive and "smarter" system in the United States by providing both near- and long-term recommendations for moving the current system into the 21st century.

## VISION FOR A "SMARTER" SYSTEM

The committee's vision for the future of OSH surveillance is a collaborative system of systems. Recognizing the varying mandates and roles of many relevant stakeholders, the committee believes that it is possible to strengthen the ongoing coordination and data sharing across federal agencies, between federal and state agencies, across state agencies (e.g., labor and health), and with employers and workers to result in the maximum possible engagement of all. A system of systems approach to OSH surveillance would minimize the undercounting of occupational injuries and illnesses by gathering sufficient data that include nontraditional occupations and worker groups in a representative manner and enhancing prevention-relevant information in surveillance data to include race and ethnicity as well as occupation and industry. It would expand outcomes to include chronic diseases and their causes and include leading indicators, primarily through adequately detailed exposure information. Further, this system would maximize appropriate use of technologies to facilitate all surveillance processes and create structures for disseminating information to levels where it can be acted upon.

Critical enabling components can be leveraged by agencies responsible for implementing the smarter system. Such efforts begin by extending the capacities of the agencies through targeted enhancements of both existing systems and technical personnel, and through effective communication across agencies. This will be complemented by information and other inputs available through employers, employee representatives, relevant intermediaries, and individuals along with engaged health care systems, all taking full advantage of state-of-the-art technology.

Engagement of employers and health care providers could result in substantive improvements by augmenting existing resources (e.g., effective implementation of electronic reporting) along with development of new resources, such as voluntary within-industry partnerships to engage collectively in exposure surveillance. Individual workers will play an essential role in the smarter system, independent of employer relationship, by participating in population health surveys that incorporate occupational information (e.g., the Household Survey of Occupational Injuries and Illness, the National Health Interview Survey, the Behavioral Risk Factor Surveillance System, and the Medical Expenditure Panel Survey). Information from these surveys complements information from other data sources as there is no single definitive data source that covers all aspects of work injuries and illnesses.

Evolving health care systems, along with technology embedded in the delivery system, can greatly facilitate enriched inputs of data on work that can be linked to health outcome data. Inclusion of occupational information in the electronic health record and advances in health care reporting structures can improve reports of work-related health conditions.

Fundamental to a successful smarter system is the sufficient and creative use of information technology capacity and resources. These include effective autocoding of occupational information in all appropriate records, electronic reporting wherever possible from all traditional and emerging reporting sources, and development of hardware and software for simplified, efficient, and real-time collection of information (e.g., exposure and compliance data). To enable the system fully, methods and tools need to be developed for timely and effective collection and analysis of surveillance data. In addition, software needs to be designed and disseminated so that all relevant stakeholders can undertake their own examination of surveillance information and act on findings as quickly as possible to improve worker safety and health.

## GETTING TO A SMARTER SYSTEM

Based on a systematic review of current surveillance efforts and barriers, the committee examined several approaches to advancing this vision. First and foremost, OSH surveillance needs to become a priority if it is to serve the core function of providing the information essential to guide public health actions to improve worker safety and health. Surveillance often exists in the background of public health programs, rising to a level of importance only at times that call for emergency action. However, the system needs to seamlessly collect, collate, and assess information without interruption to support evidence-based actions, emergency or otherwise. With surveillance as a priority, the development of a centralized coordination of a system of systems can provide the essential evidence to guide prevention efforts that advance program objectives in the most cost-effective manner.

Recommendation Q (meta-recommendation): The Secretary of HHS, with the support of the Secretary of Labor, should direct NIOSH to form and lead a coordinating entity in partnership with OSHA, BLS, and other relevant agencies. The coordinating entity should:

- develop and regularly update a national occupational safety and health surveillance strategic plan that is based on well-articulated objectives;
- coordinate the design and evaluation of an evolving national system of systems for OSH surveillance and for the dissemination of surveillance information provided by these systems;
- publish a report on progress toward the strategic plan's implementation at least every 5 years, documenting advances toward achieving a 21st Century Smarter Occupational Safety and Health (OSH) Surveillance System; and
- engage partners, including other federal health statistics agencies, state agencies with OSH responsibilities, and stakeholders.

This recommendation is the cornerstone for the advice offered by the committee. The envisioned coordinating entity is essential if the system of systems is to be a cross-agency vision and plan for moving forward, if the other recommendations are to be properly prioritized and carried out, and if the resulting system of systems is to be guided effectively by the principles and objectives of an ideal national occupational safety and health surveillance system.

As a system of systems, this smarter system will need to provide specific and clear-cut objectives for each of the surveillance systems within the overall system, and include concrete objectives for each key federal agency. It would also include detailed plans for engagement of the state-based OSH programs that identify priority conditions for expanded surveillance, provide guidance on how to use the data generated by the states, and, whenever possible, identify lead agencies (both federal and state) for these conditions. Resource needs would need to be organized, projected, and articulated. Benchmarks need to be identified and realistic timelines need to be specified for measurable progress. Furthermore, evaluation will be essential for expanding or replicating successful systems at different levels of the organization as appropriate, and for terminating systems that no longer meet objectives.

Long-term objectives will require greatly expanded data resources paired with improved methods for assessing burden. All actors face constraints due to limited resources and complicated historical precedents. Therefore, whenever possible, recommendations are presented with both near- and long-term constituents. The near-term parts are meant to be possible within existing resources and constraints. The longer-term parts are expected to require new resources (financial and personnel) along with the evolution of elements that are beyond the direct control of the leadership for OSH surveillance. Nonetheless, if the recommended stakeholders fail to take initial steps that can lead to longer-term developments, continuing progress toward the best possible surveillance system will either be delayed or impossible.

The committee reviewed evidence for assessing the magnitude of the OSH problem nationally and, where possible, at the state level. It also considered the social and economic costs of the burden of injury and illness that are borne by individuals and society. The limited evidence available still proved useful in establishing the context for the committee's work. However, neither the OSH community nor the public health community as a whole currently has adequate information and analysis to assess this burden properly and to track it effectively. Therefore, we have offered a way forward to a more complete, ongoing effort to measure and report on the burden and importance of occupational disease and injury to our nation and its people.

### The Overall System

In setting forth OHS surveillance as a national priority, the responsible agencies, most centrally NIOSH, will need to delineate a clear line of responsibility and authority over each agency's surveillance activities and personnel. Unless leadership structurally and distinctly recognizes and articulates these actions, the system is incapable of achieving the identified goals. Agencies need to ensure that links across agencies are as seamless as possible and that barriers are eliminated for timely, efficient sharing of data and information. The overall system will need to be founded on a close working relationship between federal and state partners because coordinated federal and state systems offer immense advantages over either operating independently. The report accordingly stresses the value of an effective federal-state partnership and strengthened state efforts, both to facilitate and serve a national effort to identify and monitor priority conditions and emerging problems, and to foster prevention programs at the level that can best address these concerns (see Recommendation C).

**Recommendation C: NIOSH should lead a collaborative effort with BLS, OSHA, the states, and other relevant federal agencies to establish and strengthen state-based OSH surveillance programs.**

The most effective intervention activities will need to act on the causes and not the consequences of OSH problems. Exposure and hazard surveillance points the way to primary prevention, particularly important for long-latency occupational diseases. Consequently, a sequence of efforts is designed to construct a robust exposure component of the envisioned surveillance system. The report calls for an immediate collaborative effort of federal agencies to initiate the development of a comprehensive approach for exposure surveillance that builds and updates a database of risks and exposures to predict and locate work-related acute and chronic health conditions for prevention (see Recommendation H).

**Recommendation H: NIOSH, in consultation with OSHA, should place priority on developing a comprehensive approach for exposure surveillance.**

The effectiveness of the overall system of systems will depend on the resources and methods of biomedical informatics. Developments occur so rapidly in this area that a lack of experienced, engaged personnel leads to lost opportunities and compromised system effectiveness. Most critical is the need for NIOSH to attract adequate informatics personnel and resources (see Recommendation J), though recruiting and maintaining informatics experts in the public sector can be challenging. Informatics capacity will need to be leveraged to enable OSH agencies, particularly NIOSH, to use advanced computational and analytical tools and to monitor advances in information technology (see Recommendations L and M). It will be important to engage the scientific community by working with the National Library of Medicine to facilitate easy discovery of the importance of the connections between work and disease or injury in published research (see Recommendation K). A collaborative federal effort is needed to promote and support education and training of the surveillance workforce by identifying core competencies required for OSH surveillance (e.g., epidemiology, biomedical informatics, and biostatistics) and to engage educational institutions to establish or modify training programs accordingly (see Recommendation P).

**Recommendation J: NIOSH should build and maintain a robust internal capacity in biomedical informatics applied to OSH surveillance.**

**Recommendation K: NIOSH should work with the National Library of Medicine to incorporate core OSH surveillance terminologies, including those for industry and occupation, into the Unified Medical Language System.**

**Recommendation L: NIOSH should lead efforts to establish data standards and software tools for coding and using occupational data in electronic health records.**

**Recommendation M: NIOSH and BLS, working with other relevant agencies, academic centers, and other stakeholders, should coordinate and consolidate, where possible, efforts to develop and evaluate state-of-the-art computational and analytical tools for processing free text data found in OSH surveillance records of all types.**

**Recommendation P: NIOSH, OSHA, and BLS should work together to encourage education and training of the surveillance workforce in disciplines necessary for developing and using surveillance systems, including epidemiology, biomedical informatics, and biostatistics.**

## Data Collection and Processing

Surveillance starts with collecting and processing relevant data, followed by data analysis and interpretation that can guide policy and interventions. The sources and quality of inputs to the smarter system are thus crucial. The issue of undercounting of occupational injuries and illnesses is highlighted from two perspectives: cases that are out of scope or cases that are simply unreported. Failure to count occupational injuries that are out of scope is a recognized consequence of surveys or other assessments that do not capture data on some segments of the working population. For example, the Survey of Occupational Injuries and Illnesses (SOII) does not cover or capture injuries to workers who are self-employed (e.g., independent contractors) or who work on small farms. When the 1987 NRC report called specific attention to both perspectives, the Bureau of Labor Statistics responded by addressing that which was within its control and now has ongoing efforts to understand the reasons for underreporting. The underreporting problem, however, is complex and multifactorial. Additionally, there are limitations in the identification and reporting of chronic diseases associated with work. Further, many of the systems that collect information on injuries and illnesses do not collect occupationally-related data.

One of the major inputs to OSH surveillance is through the SOII and the report discusses needed enhancements to the SOII, including that would better inform public health actions for underserved populations. Injury and illness recording can be improved by better characterizing work-related injuries and illnesses in a manner that enhances usefulness at the worksite as well as at national and state levels (see Recommendation A). Developing ways to incorporate information on race, ethnicity, and employment arrangements will allow for identification of vulnerable worker populations and risks that may be associated with different types of employment arrangements.

**Recommendation A: BLS and OSHA should collaborate to enhance injury and illness recording and the Survey of Occupational Injuries and Illnesses (SOII) to achieve more complete, accurate, and robust information on the extent, distribution, and characteristics of work-related injuries and illnesses, and affected workers, for use at the worksite and at national and state levels.**

The committee supports the BLS plan to implement a Household Survey of Occupational Injuries and Illnesses (HSOII), as it will fill in data gaps for populations of workers who are missing from employer-based injury reporting and will provide worker input (see Recommendation D). Another largely untapped resource for injury surveillance data is the workers' compensation system, and the report promotes the expanded use of workers' compensation data for occupational injury and illness surveillance (see Recommendation F).

**Recommendation D: BLS should place priority on implementing their plan for a Household Survey of Occupational Injuries and Illnesses (HSOII).**

**Recommendation F: NIOSH, with assistance from OSHA, should explore and promote the expanded use of workers' compensation data for occupational injury and illness surveillance and the development of surveillance for consequences of injury and illness outcomes, including return to work and disability.**

Work-related disease information (as opposed to injury) has been almost absent from occupational health surveillance, which was noted in 1987 and remains true today. Several recommendations address this priority component of data collection, including the enhanced assessment of self-reported health through the National Center for Health Statistics or an expanded HSOII (one component of Recommendation D) and the development of a comprehensive approach for exposure surveillance (see Recommendations D and H). Also key are occupational disease monitoring (see Recommendation B) and the specification of industry and occupation as core variables in all federal health surveys (see Recommendation G). The latter recommendation addresses the unparalleled opportunity to gain information on the distribution of exposure-related factors in a manner modeled on the highly successful experience in the European Union, which has over 25 years of experience in such efforts.

**Recommendation B: NIOSH, working with the state occupational safety and health surveillance programs and across divisions within the agency, should develop a methodology and coordinated system for surveillance of both fatal and nonfatal occupational disease using multiple data sources.**

**Recommendation G: HHS should designate industry and occupation as core demographic variables collected in federal health surveys, as well as in other relevant public health surveillance systems, and foster collaboration between NIOSH and other CDC centers in maximizing the surveillance benefits of including industry and occupation in these surveys and surveillance systems.**

For OSH surveillance, a forward-looking aspect of data collection and processing concerns how best to remove the barrier to recording and interpreting occupational information in medical records and in population surveys of all types. The report accordingly recommends that NIOSH, with an evolving biomedical informatics capacity, lead efforts to establish data standards and software tools for coding and extracting occupational data in electronic health records. These records are increasingly becoming standard practice, and there is an opportunity to make substantial long-lasting progress to eliminate barriers to linking occupation and disease (see Recommendation L). The creation of a cross-agency effort is needed to develop and evaluate state-of-the-art computational and analytical tools for processing free text data found in OSH records of all types (see Recommendation M).

## Data Analysis and Information Dissemination

Successful collection and processing of surveillance data alone does not make a successful surveillance system. The system also requires thoughtful analysis, careful interpretation, and then dissemination of results to engage in policy development or public health action for prevention. Attention to analysis and interpretation is essential when calling on partners to provide new or more data, even if that requirement is facilitated through electronic reporting. Accordingly, a program that provides for better reporting, such as the OSHA electronic reporting initiative, needs to be accompanied by a robust plan for analyzing, interpreting, and disseminating the information. OSHA, its sister agencies, and stakeholders will need to develop and publicize plans to maximize the utility of their new electronic reporting initiative by providing means and methods for ongoing analysis and dissemination of these data with special attention to serving individual employer needs while simultaneously minimizing duplication of reporting by employers (see Recommendation E).

**Recommendation E: OSHA, in conjunction with BLS, NIOSH, state agencies, and other stakeholders, should develop plans to maximize the effectiveness and utility of OSHA's new electronic reporting initiative for surveillance.**

Of equal importance is the need for the dissemination of surveillance findings and analyses in useful formats for informing and evaluating prevention. There is a need to make regular reports to the nation that publicize the overall burden of occupational injury and disease in terms of the burdens on health, the economy, and society (see Recommendation I). NIOSH, OSHA, and BLS are already engaged with dissemination and, while ideas are provided throughout the report on how these efforts could be enhanced, most important is the need for an effective alert mechanism. A smoothly operating alert mechanism should be created that receives, enhances, and rapidly publicizes to those who need to act on the signals of emerging OSH problems either as new associations of work and illness or injury or of old associations found in new settings (see Recommendation N). Appropriate and timely attention to surveillance findings, routine or new, is essential for prevention and thus requires that a smooth and centralized mechanism be established for timely ongoing dissemination of cross-agency information to all relevant actors (see Recommendation O).

**Recommendation I: NIOSH should coordinate with OSHA, BLS, and other relevant agencies to measure and report, on a regular basis, the economic and health burdens of occupational injury and disease at the national level.**

**Recommendation N: To identify emerging and serious OSH injuries, illnesses, and exposures in a timely fashion, NIOSH (in coordination with OSHA, BLS, and the states) should develop and implement a plan for routine, coordinated, rapid analysis of case-level OSH data collected by different surveillance systems, followed by the timely sharing of the findings.**

**Recommendation O: To promote and facilitate the use of surveillance information for prevention, and to present more comprehensive information on the extent, distribution, and characteristics of OSH injuries, illnesses, and exposures, NIOSH (in coordination with and input from OSHA, BLS, and the states) should establish a coordinated strategy and mechanism for timely dissemination of surveillance information.**

## CLOSING REMARKS

Worker safety and health is of paramount importance to thriving workers and workplaces, and accordingly to society as a whole. Ensuring and improving worker safety and health is a commitment taken seriously, and diligently acted upon, by numerous federal, state, and local agencies; workers and worker organizations; employers and employer organizations; and many others. More can be done to inform and improve these efforts through strengthening OSH surveillance in the United States. With the rapid changes in the nature of work in the United States, and with new risks added to those that have always existed, the nation clearly needs a smarter OSH surveillance system of systems for the 21st century. This report provides the evidence and recommendations for a greatly enhanced OSH surveillance system that is envisioned to be smarter, more dynamic, and more highly coordinated.

# 1

# Introduction

The workplace is where 156 million working adults in the United States spend many waking hours, and it has a profound influence on health and well-being. Work-related injuries and illnesses can impact quality of life (BLS, 2017a). Although some occupations and work-related activities are more hazardous than others and face higher rates of injuries, illness, disease, and fatalities, workers in all occupations face some form of work-related safety and health concerns (e.g., musculoskeletal disorders arising from repetitive motions, work-related stress, pulmonary symptoms or diseases caused by airborne exposures, and injuries resulting from falls or unstable work areas). Not only are such hazards a threat to worker well-being and hence to the nation's health, but one study estimated the annual cost of occupational injuries, illnesses, and deaths in the United States at $250 billion, which is more than the $219 billion for all cancer and approaching the $431.8 billion for all cardiovascular disease (Leigh, 2011). Reducing that health burden is the goal of occupational safety and health (OSH) surveillance.

OSH surveillance provides the data and analyses needed to improve worker safety and health and to understand the relationships between work and injuries and illnesses. Today, OSH surveillance efforts are fragmented across multiple federal and state agencies that collect, store, analyze, and interpret data to meet their specific agency needs. Agency resources dedicated specifically to work-related safety and health surveillance are limited, partly because the responsibility for occupational health and safety issues is divided between labor and public health and there has never been core public health funding allocated specifically for OSH surveillance. Many factors point to a need to rethink and improve OSH surveillance in the United States including changes in the nature of work, the workforce, and employer-worker relationships that have occurred over the past 40 years along with ever evolving methods and technologies for identifying, reducing, and eliminating work-related health and safety risks.

A more cohesive and "smarter" OSH surveillance system is needed to address the demands and concerns of the current and future workforce. It needs to be "smarter" in the sense that the system needs to be one that demonstrates efficiencies, integrates strategies across multiple data sources, coordinates efforts across key surveillance agencies, and applies domain knowledge effectively to interpret data and to deliver insights to key stakeholders. Furthermore, there is an opportunity to think creatively about envisioning how that "smarter" system might build upon the current system and regulatory climate in the United States and take full advantage of new information technologies. Additionally, effective and comprehensive OSH surveillance models from other countries can be explored. This report aims to put forth a vision and pragmatic framework for a national OSH surveillance system in the United States with the goal of improving worker safety and health.

## BACKGROUND

The topic of occupational safety and health surveillance across the United States was last examined comprehensively in a 1987 National Research Council (NRC) report *Counting Injuries and Illnesses in the Workplace: Proposals for a Better System* (NRC, 1987). In a more recent series of eight reports by the NRC and the Institute of Medicine that evaluated the National Institute for Occupational Safety and Health's (NIOSH's) research programs, all of the reports identified the need for improved OSH surveillance and for additional surveillance research (IOM and NRC, 2009). The Council of State and Territorial Epidemiologists has also issued calls for improvements to address surveillance gaps (Boulton et al., 2003; CSTE, 2010, 2014). There is recognition that surveillance could be conducted more innovatively and collaboratively; for instance, by leveraging activities and resources regardless of whether they are drawn directly from various sectors of the workforce or other public health domains, or borrowed from unrelated fields. There is also the recognition that enhanced coordination of surveillance activities could enable individual state and federal agency responsibilities to better protect workers. New data sources and information technologies, such as those supporting health care reform (the Affordable Care Act), could also be used to support occupational safety and health needs.

## Changes in the Nature of Work and the Workforce and Implications for Occupational Safety and Health

Since the 1987 report, there have been major changes in the nature of work, including employee-employment arrangements, the distribution of employment among industry sectors, and the demographics of the workforce (see Figure 1-1; BLS, 2017a). Employment in manufacturing has declined, while significant growth in employment has occurred in the service sector, including health care. Individuals are likely to be working more than one job over their working life, and may hold multiple jobs at the same time. Growth has occurred in nonstandard work arrangements, such as the use of independent contractors and the outsourcing of functions to other entities, and the development of "on-demand" or "gig" work, where employment is characterized by short-term contracts or freelance work (e.g., drivers who contract with ride-share companies) (Katz and Krueger, 2016). Furthermore, the workforce is more diverse and includes more women (Figure 1-2), racial and ethnic minorities (Figure 1-3), and immigrants (Myers et al., 2013; BLS, 2017b). With these shifts in work and the workforce, employment is more precarious and many workers lack the protections and rights afforded by laws and regulations that make them more vulnerable and subject to abuse.

Approaches to occupational safety and health surveillance have yet to evolve to address effectively the changing nature of work. For example, some current data-collection approaches that use "establishments" as the basic employer unit typically rely on single employers and worksites, which

**FIGURE 1-1** Shifts in Employment by Selected Major Industry Sectors, United States, 2006-2026. SOURCE: BLS, 2017a.

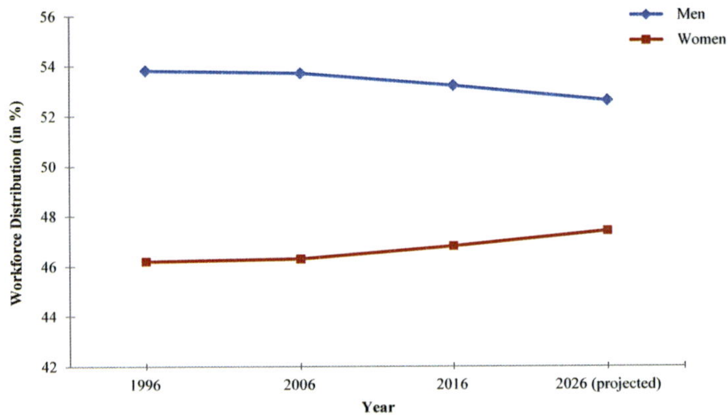


**FIGURE 1-2** Shifts in Labor Population by Gender, United States, 1996-2026. SOURCE: BLS, 2017b.

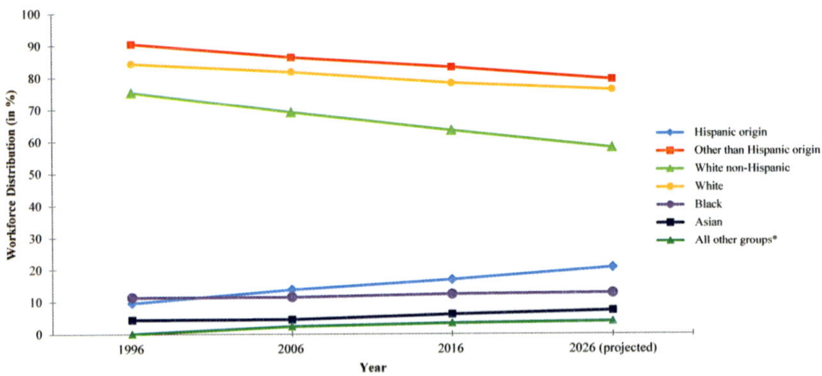


**FIGURE 1-3** Shifts in Labor Population by Race and Ethnicity, United States, 1996-2026. * The "all other groups" category includes (1) those classified as being of multiple racial origin and (2) the racial categories of (2a) American Indian and Alaska Native and (2b) Native Hawaiian and Other Pacific Islanders. SOURCE: BLS, 2017b.

are insufficient because those approaches are unable to capture gaps and emerging concerns created by the evolution of employment (Weil, 2014). Furthermore, data on hazards, protective safety, and health-management practices in individual work settings have not routinely been collected in a systematic way.

The ability to carry out rigorous workforce safety and health surveillance has other challenges as well. For example, in certain industry sectors (such as agriculture) or demographic groups (such as day laborers or teenagers), the total number of workers in an industry is necessarily a rough estimate due to temporal fluctuations or the nature of the work. The limitations of such denominator data make it difficult to determine baseline rates for injuries or illnesses in order to measure risks, track trends, and evaluate the effectiveness of efforts to reduce injury and disease.

A major change has been in the evolution, effectiveness, and relative ubiquity of information and communications technology (both methods and tools) since the 1987 report. Major advances in data collection and storage, analytic methods, sensors, and mobile devices allow information to be gathered and connected with central collection resources. Although the revolution of Internet search methods has made it easier to search for information, the quality of such large amounts of data obtained is uneven depending on factors such as source and age. Social media have also become sources of insight regarding societal trends and offer one of many methods for information dissemination that new technologies can provide. Occupational safety and health surveillance efforts will need to leverage newer technologies and tools for identifying, organizing, analyzing, and interpreting data in more innovative, powerful, and cost-effective ways. Doing so could reveal problems, trends, and emerging issues within and across sectors, groups, and geographic regions of workers. Also, these technologies offer opportunities to improve the dissemination of information to those that can use surveillance data to take preventive action, thereby improving worker safety and health and reducing associated human and economic costs of work-related injuries and illnesses.

## PURPOSE OF THIS STUDY

The National Academies of Sciences, Engineering, and Medicine formed an ad hoc committee to undertake the current study requested and sponsored by three key federal agencies involved with occupational safety and health: NIOSH, the Bureau of Labor Statistics (BLS), and the Occupational Safety and Health Administration (OSHA). The committee was tasked with developing a vision for "smarter" occupational safety and health surveillance in the United States (see Statement of Task, Box 1-1). To address the task statement, the collective expertise and experience of

**BOX 1-1**
**Statement of Task**

A more coordinated, cost-effective set of approaches for occupational safety and health surveillance is needed in the United States. A committee of the National Academies of Sciences, Engineering, and Medicine will study opportunities and provide recommendations for developing a "smarter" system.

In the course of its study, the committee will gather information about the strengths and limitations of existing national and state approaches and also review different methodologies and approaches for occupational safety and health surveillance, particularly with respect to usefulness and cost effectiveness.

Based on information gathered during the study, the committee will develop a vision for a "smarter" cost-effective occupational safety and health surveillance system; describe system components and their attributes; and recommend key steps for developing such a system. As part of its vision, the committee will

- Define essential requirements and goals for a modern occupational safety and health surveillance system; identify critical gaps to fill; reflect on how the methods, tools, and goals of surveillance may have changed since the 1987 NRC surveillance report was issued; and draw also upon other subsequent reports (for example, CDC's *Vision for Public Health Surveillance in the 21st Century*, the Council of State and Territorial Epidemiologists meeting summaries from 2009 and 2013, and other NRC reports);
- Conceptualize ways that some surveillance data might be collected, analyzed, interpreted, and disseminated more cost-effectively or innovatively (including identifying novel or underutilized means of collecting data, collecting data at different scales or different interfaces, and creating collaborations across public health and other domains), and, where possible, identify new data opportunities given current and emerging technological advancements in information technology (such as electronic health records and electronic submission of OSHA 300 logs); and
- Explore the respective current and potential roles of various federal and state agencies and private partners (such as employers and labor unions) in collecting and leveraging occupational safety and health surveillance information.

The committee will identify cost, data quality and management, and other tradeoffs inherent in different aspects of or different approaches to conducting surveillance (including the implications of using existing data systems versus collecting additional original data. It may also draw from surveillance approaches that offer insights relevant to the United States that are represented in the experience of other nations or from other fields.

The committee's recommendations will include the strengths and weaknesses of the envisioned system relative to the status quo and identify key actors (i.e., NIOSH, BLS, OSHA, etc.) and both short- and long-term actions and strategies needed to bring about a progressive evolution of the current system.

the committee includes individuals with expertise in occupational epidemiology, occupational medicine, occupational safety and health, hazard/exposure surveillance, public health, statistics, survey methods, biomedical informatics, data mining and analytics, economics, cost-benefit analysis, and workplace organization and management (see Appendix B). The committee held information-gathering meetings with invited experts to help it address its task, and members of the public were invited to attend and provide comments at these meetings (see Appendix C).

The committee intends the report to be useful to the study sponsors and to other federal and state agencies that have an interest in occupational safety and health. The report may also be of interest broadly to employers, labor unions and other worker advocacy organizations, the workers' compensation insurance industry, as well as state epidemiologists, academic researchers, and the broader public health community. The study was undertaken over an 18-month period during which there were major changes in the U.S. political scene and in government priorities. Because of uncertainties regarding those evolving priorities and resource availability, the committee has offered observations and recommendations that are intended to suggest useful efforts for both the short and long term. The report's short-term recommendations in most cases do not require new resources except in cases where new work is recommended, new resources may be required (e.g., see Recommendation D in Chapter 6). The recommendations are lettered alphabetically in the order that they appear in the report and do not correlate with their order of priority; however, actions to carry out the recommendations along with the timing and methods for addressing them will be dependent on the creation of the coordinating entity recommended in the meta-recommendation (Recommendation Q; see Appendix A for the full recommendations). The committee views the report as serving as a reference that will serve the long-term planning needs of the OSH community and other stakeholders, independent of the availability of new OSH resources or opportunities for organizational change in the short term.

## KEY TERMS AND DEFINITIONS

### Occupational Safety and Health Surveillance

Surveillance is not unique to occupational safety and health. The Centers for Disease Control and Prevention (CDC) identified surveillance as the first of the 10 core public health functions: "[To] monitor environmental and health status to identify and solve community environmental health problems" (CDC, 2011). Surveillance is defined by CDC as "the ongoing, systematic collection, analysis, and interpretation of health data, essential to the planning, implementation and evaluation of public health practice, closely integrated with the dissemination of these data to those who need to know" (Thacker et al., 2012). Many argue that surveillance is the cornerstone of public health practice, because successful prevention programs inevitably build on an effective system for monitoring and responding to what happens in the community. Quality public health surveillance efforts rely on consistent and systematic application of methods over time. Gathering data is of little value if those data are not analyzed and interpreted to produce information and understanding that is promptly disseminated to those in a position to take action.

There is a useful distinction between case- and population-based surveillance. In case-based approaches, the focus is on early identification of individual cases that require follow-up or immediate action by those who are typically legally mandated to do so (for example, the efforts of public health workers who identify and rapidly isolate or treat those with communicable diseases and of public health scientists who control or isolate an identified hazardous exposure). In population-based surveillance the focus is on a wider group and on collecting data to assess the extent of a health-related event and to monitor trends within a defined population over time and locale. Case- and population-based surveillance activities are complementary approaches. A specific circumstance may require immediate attention or intervention, even while the information about that case contributes to the ongoing collection of data about a population of similar individuals.

OSH surveillance is a subset of the broader public health surveillance field. In OSH surveillance, data are collected on work-related health outcomes (e.g., injuries, diseases, or fatalities) and on the presence of health and safety hazards (e.g., hazardous exposures or conditions). To date, the major focus has been on gathering data about health outcomes and analyzing these data to identify both sentinel cases and trends to inform the targeting and evaluation of efforts to control hazards and prevent occupational injuries and illnesses. The information produced by surveillance has also been used to inform policy development, to guide educational and regulatory activities, to develop safer technologies, and to enable research. The issue of hazard surveillance has not been emphasized in the U.S. approach to OSH surveillance, although it is a topic that may be addressed internally by some larger employers and is an area of ongoing investigation.

Research on OSH surveillance is valuable, both for deriving new insights and recommending new interventions (i.e., research "using" data collected through surveillance) and for addressing the needs of the surveillance community to develop new approaches or technologies to address shortcomings in current surveillance practices (i.e., research "in support of" surveillance processes and methods). Both surveillance activities and surveillance research are driven by the actual practice of occupational safety and health. In that applied context, information produced by the analysis and interpretation of data collected through surveillance is used to address a problem in a specific workplace, industry, or region, often working with employers, workers' organizations, and health care providers. In understanding OSH surveillance, it is useful to consider three key elements: processes that constitute a surveillance system, components that enable these processes, and methods that are applied to accomplish surveillance processes. Chapter 6 of this report focuses on the processes, and Chapter 7 identifies and discusses enabling components and associated methods (see Box 1-2 for definitions).

Issues of data completeness and data quality underlie all OSH surveillance. Throughout this report attention is directed to overcoming some of the known limitations in collecting data on the wide range of working populations. For example, one limitation in the Bureau of Labor Statistics annual Survey of Occupational Injuries and Illnesses (SOII) is that it does not include self-employed individuals, contract workers, or those in on-demand jobs.

As a core public health function, surveillance provides timely information to control or prevent exposure to hazards with the goal of reducing morbidity and mortality and improving population health. Thus, *surveillance* is generally viewed in a positive light and as a public good. Surveillance activities need to be sensitive to the privacy rights of individuals, and transparency is needed to ensure that all relevant stakeholders are informed about the steps that are taken to appropriately protect data collected about individual workers and the workplace. Thus, in public health and occupational health settings, the term *surveillance* is generally free of the negative connotations associated with government intrusion into the private lives of individuals. Concerns about privacy and confidentiality in occupational health surveillance are perhaps most pertinent if the information about an employee's health could be used by an employer against the employee. Ensuring that an ethical framework continues to undergird OSH surveillance is critical to providing necessary information to protecting workers and achieving safer and healthier workplaces.

Employers can find OSH surveillance requirements challenging. Employers understand that they have the primary responsibility for safety in their workplaces—as part of their ethical duty to their employees, as mandated by statute, and to maintain a sustainable workforce in today's economy. While employers often seek useful and timely information to improve safety in their workplaces, they may become concerned if they feel that their time and efforts are spent primarily for the sake of regulatory compliance. The OSH surveillance community is well aware of potential tensions and the need to ensure that complying with reporting requirements provide results that are valued by employers and workers.

## BOX 1-2
## Key Elements of Occupational Safety and Health Surveillance

**Processes:** These include collecting data (from employers, medical facilities, state agencies, insurance companies, and even from patients); rationalizing those data by organizing and properly curating them; sharing and merging them as appropriate; interpreting them; informing federal or other public policy; educating employers and other interested parties regarding the lessons learned; and implementing solutions, guided by policy, regulation, or naturally occurring responses to the information made available.

**Enabling Components:** The preceding listing of processes makes it clear that their implementation may be far from straightforward. The components that enable such work are both organizational and technological. They include the employers themselves, who are asked to participate actively in the collection and submission of data relevant to their company and their industry. They include health care providers who diagnose and treat health conditions and are required to report select conditions to state health agencies. They include the agencies that collect and coordinate data collection. Also relevant are other interested parties who contribute, such as employees themselves (who may experience adverse consequences by reporting on injury or illness), the trained individuals who manage the data systems, the educational programs that produce individuals (surveillance workers, employers, workers, clinicians, and others) with the necessary expertise to implement the various surveillance processes, and government oversight that seeks to enhance coordination while addressing the fiscal and organizational needs of the activity. Increasingly, however, it is the technology that enables effective and efficient implementation of surveillance processes. The devices themselves play an important role (e.g., sensors, tablet computers, electronic health records, web-based information-delivery mechanisms, smartphones, and the like), as do the developing conventions for standardizing terminology, integrating data securely in the cloud, using social media or crowd sourcing, or applying analytical software that supports the needs of the community when interpreting large amounts of collated data (Thacker et al., 2012).

**Methods:** Methods drawn from many disciplines support OSH surveillance processes and are often embedded within technological components. As noted, surveillance research has often focused on the development of such methods, whether they involve new approaches to collecting data (or new kinds of data) in the workplace, new solutions to terminology standardization and to the related encoding processes, or new analytical techniques that enhance our ability to draw conclusions from available data.

## Workplace Hazard and Exposure Surveillance

In the United States, OSH surveillance has focused primarily on health outcomes (lagging indicators) while hazard surveillance (a leading indicator) has received less attention. Hazard refers to the potential of a substance or condition to cause harm, while risk considers the probability that the hazard will cause harm. Thus, the health risk of a working condition is a function of both the presence of a hazard associated with the health condition and the extent of the exposure of an individual or population to that hazard.[1] In the context of surveillance, a hazard surveillance system (e.g., a chemical use inventory or registry) identifies potential workplace hazards and can provide information on the distribution and location of a hazard where there is potential for exposure. An exposure surveillance system, monitors actual contact that workers have with the substance or condition (potentially including data on who the workers are who were exposed, the duration of the exposure, and the level or intensity of the exposure).

Although the discussion above is most easily understood in terms of airborne chemical hazards or physical agents, it is equally relevant to acute injury risks. For instance, work at height may be considered a hazard, whereas the exposure would be the proximity of the work to an unguarded leading edge. Similarly, moving heavy objects to an elevated position may constitute a hazard, the exposure would be related to the weight, the frequency and the biomechanical load determined by the lifting task, and presence of mechanical lifting aids.

The relationships among hazard, exposure, and the "work-relatedness" of an illness or injury can be highly complex and subject to interpretation (Oppliger and Seixas, 2017). An injury or illness is work related if it was sustained in the workplace. However, only a small proportion of illnesses where work is a component in the mix of causes are currently ascribed to work exposures. For example, an individual's lung disease (e.g., chronic obstructive pulmonary disease or asthma) or chronic back strain can be due to work-related exposures along with other factors in the worker's life or genetic makeup (see Chapter 4). Oppliger and Seixas (2017) note that "Even ascribing acute injury to work is not as obvious as identifying the location at which the injury occurs; an injury at work may be partially due to non-work-related stressors (e.g., personal stressors affecting attention or vigilance) while injuries off the job may commonly have work-related contributions (e.g., neurologic toxins, long hours, shift work, or other stressors causing fatigue, etc.)". Thus, for both acute and chronic injuries and illnesses, agreed-upon surveillance case definitions are necessary to

[1] Risk=f(Hazard, Exposure). Where risk is the probability of an adverse outcome in a defined population; hazard is the inherent damaging potential of agent or condition; and exposure is the probability, frequency, and intensity of the agent interacting with an individual.

manage the uncertainty of the contribution of work to a wide range of health conditions.

In addition to physical, chemical, and biological agents, musculoskeletal strain is increasingly recognized in many jobs associated with a wide range of both acute and chronic injury and a significant degree of disability. Even so, the assessment of various elements of exposure for musculoskeletal risks—including force, posture, repetition, work pace, work-rest cycles, and sedentary work—continues to present significant challenges.

The assessment of hazardous psychosocial exposures in the workplace is also challenging, with the potential for health impacts related to work organization (e.g., numbers of coworkers, social support, reporting relationships, and supervisory responsibilities), job demands (e.g., hours, shift work, unpredictability, and exposure to demanding customers), job control (e.g., ability to control rate or order of work components), and fair treatment (e.g., perceptions that both the individual and workers in general are being treated fairly by the employer). Health effects generally associated with these include cardiovascular disease, depression and anxiety, suicide, effects related to sleep deprivation including injury risk, and even workplace violence.

### Costs, Effectiveness, Efficiency, and Cost Effectiveness

One important goal of a 21st-century OSH surveillance system is to collect the most relevant data at the lowest cost feasible. A surveillance system would meet the definition of being *cost effective* if it identifies data with the biggest impact on improving worker safety and health while minimizing the cost of collecting those data (Gold et al., 1996; Haddix et al., 1996). A cost-effective surveillance system would also recognize trade-offs between the importance of individual data elements in identifying factors that can be modified to improve worker safety and health and the costs of collecting each of those data elements. As a result, a cost-effective surveillance system might be less comprehensive than an ideal system because the costs of collecting the ideal set of surveillance data might be judged to be prohibitively expensive from a societal perspective. Finally, an *efficient* system for collecting OSH data may not be effective or cost effective if it targets the collection of data that have only minimal impact on improving worker safety and health outcomes.

A fundamental concept in conducting a cost-effectiveness analysis of any intervention, including an OSH surveillance system, is that costs and benefits are always measured and compared from a societal perspective (Gold et al., 1996; Haddix et al., 1996; Muennig, 2008). A societal perspective is preferred when considering the value of public investments and policy because comparing costs and benefits solely from the perspective of individual stakeholders can lead to erroneous conclusions and recommendations. Thus, this report adopts a societal perspective of cost effectiveness.

The costs of a surveillance system include not only the costs of data collection, but also the costs associated with the health and productivity consequences of occupational exposures, injuries, illnesses, and mortality on workers, their families, and society. As previously noted, the most recent estimate of the total economic burden of occupational illness, injury, and death is $250 billion annually in the United States (Leigh, 2011). Therefore, an improved OSH data surveillance system can be used by policy makers and stakeholders to prioritize interventions for reducing the economic costs of occupational illness, injury, and death to society. Identifying areas of greatest need for OSH interventions through use of an improved surveillance system and then targeting effective interventions based on previous research and evidence is likely to produce significant and substantial savings to employers, employees, and society (Schulte et al., 2017). The importance of assessing costs and benefits of an improved OSH surveillance system is discussed further in Chapter 7.

## ORGANIZATION OF THE REPORT

This report covers the breadth of the committee's statement of task. Chapter 2 sets forth the guiding principles and objectives developed by the committee and used as a basis for the report's recommendations. The chapter also provides an overview of the committee's vision of a future "smarter" surveillance system. In Chapter 3 the committee examines the federal and state agencies and stakeholders engaged in occupational safety and health surveillance in the United States. It also summarizes the major recommendations of the 1987 NRC report and the response to that report's recommendations. Chapter 4 provides a more detailed summary of the current status of OSH surveillance in the United States. It summarizes the current approaches to surveillance of fatal and nonfatal occupational injuries and of occupational illnesses. The chapter also discusses the current, albeit limited, state of hazard surveillance and addresses the crosscutting issues of state-based surveillance, surveillance research, and data on populations at risk. OSH surveillance activities outside the United States are summarized in Chapter 5, emphasizing activities that may offer lessons for ongoing work in this country. To address the gaps identified in Chapter 4, Chapter 6 then discusses the promising developments, processes, and technologies that can be used to improve OSH surveillance, including electronic reporting initiatives, electronic health records, the mobilization of workers' compensation systems, and new directions for hazard surveillance. This is followed by a focus in Chapter 7 on key actions essential for moving forward with an improved national OSH surveillance system, emphasizing (1) quantification of the economic and health burden of occupational illness and injury, (2) coordination of surveillance efforts, (3) use of information technology, and (4) enhanced training and support for surveillance practitioners. The report's concluding chapter offers a final overarching meta-recommendation and summarizes the report's conclusions and recommendations for achieving a smarter OSH surveillance system.

## REFERENCES

BLS (Bureau of Labor Statistics). 2017a. *Employment by major industry sector*. Available online at https://www.bls.gov/emp/ep_table_201.htm (accessed November 21, 2017).

BLS. 2017b. *Civilian labor force participation rate, by age, sex, race, and ethnicity*. Available online at https://www.bls.gov/emp/ep_table_303.htm (accessed November 21, 2017).

Boulton, M. L., R. A. Malouin, K. Hodge, and L. Robinson. 2003. Assessment of the epidemiologic capacity in state and territorial health departments—United States, 2001. *MMWR* 52(43):1049-1051.

CDC (Centers for Disease Control and Prevention). 2011. *Core functions of public health and how they relate to the 10 essential services*. Available online at https://www.cdc.gov/nceh/ehs/ephli/core_ess.htm (accessed April 10, 2017).

CSTE (Council of State and Territorial Epidemiologists). 2010. *2009 National assessment of epidemiology capacity: Findings and recommendations*. Available online at http://www.cste2.org/webpdfs/ecabrieffinal2010.pdf (accessed August 22, 2017).

CSTE. 2014. *2013 National assessment of epidemiology capacity: Findings and recommendations*. Available online at http://www.cste2.org/2013eca/CSTEEpidemiologyCapacityAssessment2014-final2.pdf (accessed August 22, 2017).

Gold, M. R., J. E. Siegel, L. B. Russell, and M. C. Weinstein. 1996. *Cost-Effectiveness in Health and Medicine*. New York: Oxford University Press.

Haddix, A. C., S. M. Teutsch, P. A. Shaffer, and D. O. Dunet. 1996. *Prevention Effectiveness: A Guide to Decision Analysis and Economic Evaluation*. New York: Oxford University Press.

IOM and NRC (Institute of Medicine and National Research Council). 2009. *Evaluating Occupational Health and Safety Research Programs: Framework and Next Steps*. Washington, DC: The National Academies Press.

Katz, L. F., and A. B. Krueger. 2016. *The rise and nature of alternative work arrangements in the United States, 1995-2015*. National Bureau of Economic Research: NBER Working Paper No. 22667. Available online at http://www.nber.org/papers/w22667 (accessed November 29, 2017).

Leigh, J. P. 2011. Economic burden of occupational injury and illness in the United States. *Milbank Quarterly* 89(4):728-772.

Muennig, P. 2008. *Cost-Effectiveness Analyses in Health: A Practical Approach*, 2nd ed. San Francisco, CA: Jossey-Bass.

Myers, D., S. Levy, and J. Pitkin. 2013. *The Contributions of Immigrants and Their Children to the American Workforce and Jobs of the Future*. Washington, DC: Center for American Progress. Available online at: https://www.americanprogress.org/issues/immigration/report/2013/06/19/66891/the-contributions-of-immigrants-and-their-children-to-the american-workforce-and-jobs-of-the-future/ (accessed November 22, 2017).

NRC (National Research Council). 1987. *Counting Injuries and Illnesses in the Workplace: Proposals for a Better System*. Washington, DC: National Academy Press.

Oppliger, A., and N. Seixas. 2017. What makes a disease 'occupational'? *Annals of Work Exposures and Health* 61(2):135-136.

Schulte, P. A., R. Pana-Cryan, T. Schnorr, A. L. Schill, R. Guerin, S. Felknor, and G. R. Wagner. 2017. An approach to assess the burden of work-related injury, disease, and distress. *American Journal of Public Health* 107(7):1051-1057.

Thacker, S. B., J. R. Qualters, and L. M. Lee. 2012. Public health surveillance in the United States: Evolution and challenges. *Morbidity and Mortality Weekly Report* 61(3):3-9.

Weil, D. 2014. *The Fissured Workplace: Why Work Became So Bad for So Many and What Can Be Done to Improve It.* Cambridge, MA: Harvard University Press.

# 2

# Building a "Smarter" National Surveillance System

To move forward in improving the health and safety of workers through a national occupational safety and health (OSH) surveillance system, the committee developed guiding principles, objectives, and a narrative overview of an ideal surveillance system. As discussed in this chapter, the committee recognizes that an ideal system will take the commitment and resources of numerous stakeholders including workers, employers, health care providers (primary care, occupational health, public health), and regulatory, public health, policy, and research agencies at the local, state, and national levels.

A starting point in this discussion is the definition of public health surveillance as the "ongoing, systematic collection, analysis, and interpretation of health data, essential to the planning, implementation and evaluation of public health practice, closely integrated with the dissemination of these data to those who need to know" (Thacker et al., 2012). To characterize surveillance further: *Systematic* refers to using consistent methods over time and *ongoing* refers to continuous or periodic collection of data useful to identify patterns and trends through *analysis*, which entails routine and targeted application of statistical methods to data, producing information, which public health practitioners *interpret* to understand the possible need for public health action. The final links in the surveillance chain are the actual communication of important information to those responsible for taking action and the application of data to prevention (Thacker et al., 2012).

## GUIDING PRINCIPLES

In setting forth the vision for a national OSH surveillance system to improve worker safety and health, the committee established a set of guiding principles:

*Guiding Principle 1—Sustain strong leadership:* Robust and collaborative federal leadership built on strong ties with states and other relevant stakeholders is critical to successful occupational safety and health surveillance. Engagement of the community of users who need occupational safety and health surveillance information for action is essential.

*Guiding Principle 2—Ensure quality data:* Continuous monitoring of data quality and program activities is essential to ensure program efficiency and impact.

*Guiding Principle 3—Protect data:* Privacy, confidentiality, and access to data would be safeguarded while maximizing the utility of surveillance information for prevention.

*Guiding Principle 4—Disseminate widely:* Timely analysis, and interpretation of surveillance inputs with routine dissemination of information in relevant formats, promotes the use of surveillance information for action by all stakeholders.

*Guiding Principle 5—Support the surveillance workforce:* An efficient, reliable national surveillance system requires public health professionals with training in occupational injury, illness, and hazard surveillance, and with the tools and technology necessary to achieve surveillance objectives.

*Guiding Principle 6—Set and follow effective standards:* The consistent use of standards for data collection, analysis, and information presentation and dissemination will heighten the efficiency and effectiveness of OSH surveillance.

## OBJECTIVES

The overarching aim of an OSH surveillance system is to protect the health of working people by providing the information needed to target, design, and evaluate efforts to control workplace hazards, prevent occupational injuries and disease, and reduce associated societal costs. To serve this aim, an ideal national system would have the following objectives:

**Objective 1: Guide immediate action to control threats to occupational health and safety.**

Capacity is needed to identify health-related events that require immediate control efforts and to provide data and analysis in service of those efforts. Further, capacity is needed to prevent continued exposure to the identified risk. The speed of response is conditioned on the nature of the health event and on the need to identify the cause, the vulnerability of the population at risk, and the length of the latency judged to exist between exposure and health outcome. National surveillance can facilitate the identification of situations requiring immediate response, which will then require interpretation and action not only by national agencies but also by state and local agencies, employers, workers, and others.

**Objective 2: Measure the burden of work-related injuries or illnesses and monitor trends over time and space.**

The many measures of the burden of work-related injury and illness include the type of injuries and illnesses; the number and rate of injuries and illnesses; indicators of severity, such as requirements for hospitalization, surgery, or number of lost workdays; the short- and long-term economic and social costs; and estimates of the burden of preventable disability, such as disability-adjusted life years, and years of life lost. Monitoring trends in these measures consistently at a national scale allows for assessment of the progress toward national occupational health and safety goals and to the identification of unfavorable trends that require further attention. Further characterizing the burden by geographic region, by time period, by the nature and cause of injury or illness, and by the presence of known risk factors will allow intervention efforts to be targeted more efficiently.

**Objective 3: Identify industries, occupations, and worksites as well as populations, defined by sociodemographic characteristics or work arrangements, at high risk for work-related injury, illness, or hazardous exposures.**

Hazardous workplace exposures, work-related injuries, and illness are not distributed equally across the workforce or workplaces. Characterizing the impact of work-related injury and illness across working populations permits attention to be directed to populations at higher risk. Working populations are usually described through the demographics of the workforce—age, gender, race and ethnicity, language preference, nativity, or by work and exposure characteristics defined by industry and occupation of employment, job task or specific chemical, and physical or biological workplace exposures. Other characteristics of work directly related to occupational health include work patterns such as shift work, seasonal work, as well as employer and workplace characteristics such as self-employment, employee turnover rates, and fixed- and non-fixed-location workplaces. Newer working arrangements add important complexity and will require creative efforts for surveillance systems to adapt collection procedures to describe these work settings, particularly multiemployer worksites and work arrangements that include temporary work either through temp agency hires or direct-hire temps and "gig" economy workers.

**Objective 4: Detect and respond to new or emerging workplace hazards or facilitate the investigation of new diseases linked to occupational exposures.**

The capacity for identifying new and emerging hazards lies largely in case-based surveillance systems involving reporting by employers, employees, and medical entities such as hospitals, laboratories, or health care providers. Ongoing examination of patterns evident in population-based surveillance efforts also can point to emerging problems. The reporting systems and infrastructure to collect and analyze these data require collaborations between state and federal OSH surveillance programs as well as the capacity to rapidly and effectively disseminate information on suspected new causal relationships that impact worker health. Most obviously, this is a concern for new industrial applications but experience suggests that known hazards continue to appear in new settings that can remain hidden if surveillance is not effectively implemented and the data generated and analyzed.

**Objective 5: Guide the planning, implementation, and evaluation of programs and policies intended to prevent and control work-related injuries, illnesses, and hazardous workplace exposures.**

Surveillance activities and data at the establishment level and at local, state, or national levels provide essential information upon which to build and implement effective occupational health and safety prevention programs. Once implemented, these same surveillance activities allow ongoing assessment of success as well as highlighting where additional efforts are needed. There are a wide variety of programs, each of which requires thoughtful and innovative use of surveillance data to track impact. Examples include those targeting workers (e.g., worker training programs), management administrative efforts (e.g., health and safety management programs), technology impacts (introduction and effective functioning of

**TABLE 2-1** Users and Uses of OSH Surveillance Information

| Users | Examples of Uses of Surveillance Information |
|---|---|
| Federal and state regulatory and enforcement agencies | Standard setting, enforcement targeting, priority setting and program evaluation, evaluating state workers' compensation insurance programs, and workers' compensation insurers |
| Federal and state health agencies | Research, conducting population- and case-based surveillance, identifying high-risk populations and worksites, identifying emerging work-associated problems, identifying hazardous conditions and exposures, community health needs planning, and targeting and evaluating interventions |
| Federal and state policy makers | Assessing the effectiveness of laws, regulations, and programs; identifying problems and unmet needs; and setting priorities for funding and program activities |
| Workers' compensation insurers, insurance community, and attorneys | Advance risk management and loss control activities; influence premium setting; allocate liability and work-relatedness |
| Employees and unions | Needed at the worksite, industry, and occupational levels for identifying hazardous conditions, jobs exposures, and hazards at the worksite and with the occupation; establish safety priorities; focus training programs; focus advocacy efforts |
| Employers and employer organizations | Needed at the worksite, industry, and occupational levels for identifying hazardous conditions, jobs exposures, and hazards at the worksite and with the occupation; establish safety priorities; focus on training programs; focus on advocacy efforts |
| Clinicians | Needed for the diagnosis and management of occupational disease; characterizing exposures to physical, chemical, and biological hazards at work; and determining light duty and safe return to work |
| Researchers | Conducting research on the relationship between exposures and hazards and injuries and illnesses; effectiveness of control measures; assessing the economic and social impacts of injuries, illnesses, and deaths; determining predictors of return to work; and compiling background data for needs assessment and project justification in research applications |
| Media/communicators | Research and background for stories on significant and emerging hazards, regulatory and enforcement actions, and practices of particular employers or industries |
| Community organizations | Conducting needs assessment, setting priorities, and supporting targeting interventions (e.g., by coalitions for occupational safety and health worker centers) |
| Equipment manufacturers/ process/product designers | Identify requirements for equipment and technology changes, and prevention though design options |

specific and company-wide engineering controls), health care personnel and systems (e.g., medical monitoring for consequences of work exposures), employer workplace prevention initiatives (e.g., employer confidentially benchmarking individual establishment performance against industry overall performance levels), and surveillance programs (e.g., environmental monitoring and patterns of illness or injury at work, effectiveness of regulations, success of industry, or geographic-wide risk control efforts).

**Objective 6: Generate hypotheses and make data available for research.**

In addition to immediate public health action, surveillance data offer great value in enabling new understandings of OSH risks and prevention opportunities. Surveillance can identify unusual patterns of disease or injury or unanticipated associations that indicate the need for further investigation. Follow-up research that takes full advantage of surveillance data can yield additional information about potential risk factors for the health outcomes or exposures under surveillance and generating hypotheses stimulating further research. An overview of the range of different users and associated uses for surveillance findings arrayed across the surveillance objectives is provided in Table 2-1.

## AN IDEAL SURVEILLANCE SYSTEM

The guiding principles and surveillance system objectives outlined above serve as the basis for the report's recommendations. The following overview also provides a description of the committee's vision for an ideal national occupational health surveillance system for the United States—a model to consider as the proposed surveillance system evolves.

The ideal "smart" national OSH surveillance system is best thought of as not a single system but a system of systems that incorporates and coordinates the collection and analysis of a variety of data sources for capturing workplace exposures and hazards, work-related injuries and illnesses, and their determinants and then provides the platforms and coordinating tools to use the information to improve worker safety and health. The ideal system:

- Relies on dedicated occupational injury and illness surveillance systems as well as inclusion of occupational information into other public health surveillance systems;
- Maximizes the use of new information technologies and changes in health care systems to enhance the efficiency of data collection; provide the flexibility to change the data parameters collected and ensure consistency in data coding; and ensure data quality and timeliness;

- Ensures that an individual workers' health data are kept confidential while maximizing the utility of these data for prevention; and
- Meets the core surveillance objectives at the national, state, and local levels, is periodically reviewed, and, as appropriate, is modified. Oversight would include prioritizing and coordinating surveillance programs and surveillance research and support activities across the federal agencies. The oversight would provide direction to surveillance activity located within federal agencies charged with conducting surveillance.

The envisioned ideal system is responsive and flexible. It is constructed to take advantage of existing data systems where possible, while initiating efforts to establish and support new data systems to fill identified gaps. To enable linkages across systems and to provide a snapshot of the population's occupational safety and health status, data-collection and data-management efforts are standardized. It is also coordinated with surveillance activities across all of the Centers for Disease Control and Prevention (CDC) and their respective state and local partners, to enable and promote consideration of occupational health concerns (i.e., the impact of work on health) within the broader public health context.

The populations under surveillance include all currently employed, disabled, and retired individuals in the United States, regardless of the characteristics of their employment arrangements. Characteristics of the population to be collected include age, gender, race and ethnicity, socioeconomic status, shift status, duration of employment, work status, employment arrangement, and security. It would include the collection of information that would allow for the identification of vulnerable populations and occupational health inequities across racial and ethnic groups. Characteristics of workplaces to be collected include size, multiemployer site, public or private status, and union status to identify the extent to which these factors can direct attention to the risks for injury and illness that need attention.

The system involves collaboration and information sharing between labor, public health, and multiple agencies at all levels of government, including the Occupational Safety and Health Administration (OSHA), the Mine Safety and Health Administration (MSHA), the National Institute for Occupational Safety and Health (NIOSH) and other components of CDC, the Department of Transportation, the Federal Aviation Administration, the Department of Energy, the Federal Railroad Administration, and the Nuclear Regulatory Commission. It fosters collaboration across public health domains, allowing for leveraging both surveillance and intervention resources across all relevant domains (e.g., transportation safety, violence prevention, and infectious disease control). The system likewise involves collaboration with a wide range of partners—health care professionals and providers, insurers, industry, and labor—responsible for generating health data and using surveillance findings to improve worker safety and health. It allows for ongoing input from data users and provides timely and sometimes "real-time" data in formats that meet their needs. Periodic comprehensive reports would be compiled and disseminated that combine information from multiple sources (i.e., employer and employee surveys for injuries and surveys and medical records for illnesses) to ensure the most comprehensive information that minimizes the possibility of providing only a partial picture of the burden of work-related injuries and illnesses.

## TOOLS AND SOURCES FOR DATA COLLECTION

Very broadly, the data needed for a national occupational injury, illness, and hazardous exposure surveillance system falls into linked content areas—worker health data linked to the worker's employer and type of work data. In an ideal system, worker health data span from birth through death linking socioeconomic, environmental, and behavioral influences on health with the worker's personal health records. Worker employment data characterize (and where possible quantify) exposures to chemical, physical, biological, and psychosocial hazards and employment security throughout the worker's employment history.

In an ideal system, the national OSH surveillance structure has multiple data-collection processes, which can operate separately or together to build surveillance systems that are needed by the nation. The system has multiple data-collection mechanisms (e.g., data from emergency departments, death certificates, hospital medical records, poison control centers, population-based surveys, hazard data collection, medical laboratories, and specialized radiograph interpretations) such that the nation's surveillance system can adapt to emerging workplace safety and health issues, rely on data at established periodic intervals, and be assured of data quality and consistency across data-collection systems. Reflecting the advancing knowledge of how work impacts overall health, all health data-collection systems outside the purview of the occupational safety and health surveillance community would include information characterizing work.

Integrated data-collection systems for OHS surveillance consist of deploying worker and employer surveys, using medical and biological testing results, accessing and using administrative data, mandating reporting of specific diseases or exposures, and using vital statistics databases. Often these systems are supplemented with less traditional data sources to assign the outcome as being work related or to incorporate other content that may be valuable for prevention. For example, newspaper articles, police or coroner reports, and other data sources are used in the determination of the work-relatedness of a fatality supplementing higher value data on the death certificate maintained by the state vital statistics program. Likewise, data collected from equipment failures or manufacturing deficiencies may supplement information about injury causation. The ideal system has real-time data submitted electronically to minimize the cost and time involved in reporting and cost and time to compile the data and make the data available to stakeholders. Data collected include elements identifying an injury as associated with work independent of payer information (i.e., workers' compensation). Data on the medical and surgical treatment of work injuries assist in describing injury sequelae and evaluating best practices for treatment, predicting return to work and disability outcomes. Using the medical record and/or workers' compensation payment records, the injury and current treating diagnoses are periodically described (e.g., at 1 month and at recovery) and coupled with employment records to determine predictors for poor outcomes including return to work. Severity data, as measured through hospitalization, surgery, medication use, medical costs, time loss duration, and measures describing the injury linked to the worker's economic and social well-being, are collected from access to the medical record and from injured worker surveys.

## DATA CURATION

National surveillance data will rest within federal agencies that have the responsibility not only to initiate public health actions in a timely manner but to act in a role as data curator—acquiring record-level data, maintaining data quality, and sharing data in a usable form over the lifespan of these data—thus maximizing the use of these data for public benefit. For this to happen, all the involved agencies need to have staff with the education and training to perform this data curation.

With the adoption and execution of policies for the federal government to procure sponsored research data and create public and researcher access to these data, federal agencies sponsoring data collection for surveillance, research, or other activities will acquire significant quantities of digital data. As such, data-collection systems designed for the surveillance of specific occupational conditions or exposures will live beyond their initial and intended usage.

Digital curation is "the active management and enhancement of digital information assets for current and future use" (NRC, 2015). For data to be useful, they need to be accompanied by documentation of their production, processing and analysis workflows, and their technical structure and format. Our ideal national surveillance system relies on efforts to capture occupational safety and health data in a standardized manner across state, federal, and all other data-collection systems. Standardized approaches to collecting information about work and employment characteristics (e.g., industry and occupation), using tools to transform unstructured data to more structured data (e.g., autocoding mechanism of injury), and methods to impute data of value (e.g., race and ethnicity) are used consistently in surveillance programs. Given the importance of work as a determinant of health, all deliberations regarding the development and evolution of data-collection systems (e.g., the electronic health record or redesign of the National Health Interview Survey) will include members of the occupational safety and health research community.

Occupational safety and health digital data curation is sponsored by NIOSH with allocated staff and information technology resources with expertise in the field (NRC, 2015). Data from internal NIOSH projects, from NIOSH cooperative projects, and from other data holders such as OSHA, BLS, and state workers' compensation programs would be submitted to a digital data repository. Data curation is not universally applied to all data but considers the long-term utility of the data for research and prevention purposes. Surveillance systems capturing unique content, or those requiring large expenditures of resources, such that duplication is unlikely, are stored. All data, metadata, and analysis methods are documented and provided to the data depository. Legal and ethical constraints regarding making the data publicly available are adhered to yet the agencies balance confidentiality with the interests of protecting workers from worker injury or illness. Systematic efforts to build data repositories which provide record-level individually identifiable data to qualified researchers are enabled through legislation and policy adoption.

## BARRIERS TO A SMARTER SURVEILLANCE SYSTEM

There are numerous impediments to improving the current occupational health surveillance system, let alone to developing an ideal system. The committee has grouped these impediments into four categories: (1) confidentiality/privacy; (2) cost; (3) expertise and workforce structure; and (4) culture and mission of organizations. Box 2-1 illustrates the key barriers

**BOX 2-1**
**Key Barriers to a Smarter Surveillance System**

**Confidentiality/ Privacy**

- Employer concerns about the availability of facility data
- Employee concerns about consequences of reporting a work-related injury or illness
- Protecting confidentiality while maximizing use of the data

**BOX 2-1 Continued**

- Grouped data publication rules
- Using individual case based reports to initiate follow back action
- Merging individual case reports across multiple data sets

**Cost**

- New worker survey proposed by BLS
- Expanding information collected by BLS in the SOII
  - Collection of existing exposure data
  - Generation of exposure data
- Minimum state-based occupational disease and injury surveillance at a state agency in all 50 states
- Multisource surveillance of priority occupational injuries and illnesses in a limited number of states
- Collection of information on injury and illness treatment and management
- Adding information on occupation and industry into existing surveys/databases

**Expertise and Workforce Structure**

- Adequate epidemiologic expertise at OSHA to use surveillance data for targeting enforcement and educational intervention
- Adequate informatics capacity at NIOSH to advance an approach for collecting occupation/industry/work-relatedness in electronic medical records, autocoding of occupation/industry and other software tools to enhance surveillance efficiency

**Culture and Mission of Organizations**

- Requirements and culture at BLS for confidentiality
- Emphasis at NIOSH on research and less commitment to surveillance than other parts of CDC
- Lack of data standardization to manage occupational information across data systems
- Methods development to utilize exposure data
- Auto coding systems to manage occupational information across data systems
- Uniqueness of state-based worker compensation agencies
- Organization of CDC and lack of appreciation of the effect workplace exposures have by public health practitioners that impedes the interaction of individuals' addressing work-related health issues/exposures with individuals at CDC addressing chronic and communicable diseases
- Lack of demand for occupation/industry/exposure information in standardized electronic medical records by clinicians and institutions providing health care, which leads to a lack of emphasis on occupation/industry/exposure by agencies and vendors responsible for standardizing electronic medical records and developing software for inclusion of occupational information

within each of these four categories. These barriers lead to multiple gaps in what data are collected and effective use of the data in the current occupational health surveillance system. These gaps are discussed in Chapters 3 and 4. The committee has provided recommendations that are considered feasible to address many of the major gaps identified in Chapters 3 and 4 but understands that to overcome the key barriers identified, particularly related to cost and agency culture, will require a commitment and prioritization by stakeholders and multiple agencies at the federal and state level.

## REFERENCES

NRC (National Research Council). 2015. *Preparing the Workforce for Digital Curation.* Washington, DC: The National Academies Press.

Thacker, S. B., J. R. Qualters, and L. M. Lee. 2012. Public health surveillance in the United States: Evolution and challenges. *Morbidity and Mortality Weekly Report* 61(3):3-9.

# 3

# Overview of Agencies and Stakeholders

Occupational safety and health (OSH) surveillance is a collaborative effort of federal, state, and local agencies and stakeholders across employers, employee organizations, professional associations, and other organizations. The federal agencies that play the major roles in OSH surveillance are the Bureau of Labor Statistics (BLS), the Occupational Safety and Health Administration (OSHA), and the Mine Safety and Health Administration (MSHA) in the Department of Labor, and the National Institute for Occupational Safety and Health (NIOSH) in the Department of Health and Human Services (HHS). In addition, there are a number of other federal agencies with responsibilities and programs pertaining to OSH surveillance and prevention. State agencies also play a critical and complementary role in partnership with federal agencies. State agencies collect, analyze, and disseminate data from local sources to guide preventive action at the state, regional, and local levels; provide data to federal agencies to be aggregated for national surveillance; and fill in gaps in national surveillance data. The strong role of workers and employers is key to ensuring accurate and complete data and to using this information to implement improvements in worker safety and health at the workplace. In addition, health care facilities and organizations, workers' compensation systems, and insurance companies have data that are relevant to occupational safety and health.

This chapter's overview highlights the varying roles that different agencies and other stakeholders fill and provides background for discussions throughout the report on integrating and coordinating the multiple surveillance databases and activities that span numerous sources.[1] Additionally, the chapter provides a summary of the key recommendations made in the 1987 National Research Council (NRC) report, *Counting Injuries and Illnesses in the Workplace: Proposals for a Better System*, and points to an overview of progress to date on those recommendations (Appendix D).

## BUREAU OF LABOR STATISTICS

BLS in the Department of Labor has responsibilities for extensive data collection and analyses including statistical assessments of employment, unemployment, pay and benefits, inflation and prices, productivity, workplace injuries and illnesses, and consumer expenditures across the U.S. economy. Established in 1884 as an agency of the Department of the Interior, BLS has a long history of collecting statistical data on work-related injuries, illnesses, and fatalities, dating back to the late 1890s predating the Department of Labor. Early BLS reports in this area focused primarily on studies of individual industries, such as a 1909 report on phosphorous poisoning in the match-making industry (Drudi, 2015).

The inadequacies in workplace health and safety statistics and the need for an accurate uniform data reporting system for work-related injuries and illnesses were well recognized in the development of the Occupational Safety and Health Act of 1970 (P.L. 91-596; OSH Act). There was a general consensus that pertinent and reliable information was a prerequisite for effective health and safety programs. Section 24 of the OSH Act directed the Secretary of Labor, in consultation with the Secretary of Health and Human Services, to "develop and maintain an effective program of collection, compilation and analysis of OSH statistics." These statistics were to include "all disabling, serious, or significant injuries and illnesses, whether or not involving loss of time from work, other than minor injuries requiring only first aid treatment and which do not involve medical treatment, loss of consciousness, restriction of work or motion, or transfer to another job."

The Secretary of Labor delegated responsibility for this activity to BLS. Beginning in 1971, BLS implemented an annual nationwide survey (the Survey of Occupational Injuries and Illnesses [SOII]) of a sample of private-sector employers to collect data on occupational injuries and illnesses (NRC, 1987). MSHA and the Federal Railroad Administration provide data on mining and railroad employees. As part of a Supplementary Data System, BLS also collected data on injuries, illnesses, and demographics

[1]The committee's review and report focused on OSH surveillance programs that pertain to the civilian workforce. The Department of Defense and military branches also have surveillance programs for military personnel, but those were not addressed in this report. A review and examination of the military's surveillance programs could be considered in the future.

from 27 state workers' compensation systems. The 1987 NRC review of these systems recommended modification of the Supplementary Data System given concerns about the accuracy and generalizability to the nation and the collection of more detailed data in the SOII (NRC, 1987). BLS subsequently discontinued the Supplementary Data System. In 1992, BLS implemented changes in the SOII by collecting more detailed information on injury and illness cases resulting in one or more days away from work. Two studies, one in New Jersey and one in Texas, performed at the request of the 1987 NRC committee, found that the BLS annual survey missed 50 percent of the acute traumatic work-related fatalities reported by employers. To address the undercount in fatalities, BLS in 1992 also implemented the Census of Fatal Occupational Injuries (CFOI). Both of these nationwide surveillance systems are population-based systems administered by BLS, working in collaboration with state agencies, most often state labor departments, that share the costs. CFOI and SOII are further described in Chapter 4.

BLS has a number of additional OSH surveillance initiatives that include, among others, examining the extent and factors contributing to undercounting in the SOII, exploring the feasibility of a nationwide household survey on nonfatal occupational injuries and illnesses, and developing electronic tools for assigning standardized codes to narrative text information in occupational health and safety data sources (see further details in Chapters 4 and 6). BLS also conducts important research on OSH topics, such as an examination of workplace violence against psychiatric aids and technicians (BLS, 2015) and workplace injuries from falls (BLS, 2016a). BLS assists external researchers, providing access to SOII and CFOI microdata under strict conditions of confidentiality, as discussed below.

It has been BLS's longstanding policy and practice to pledge to respondents that it will treat the microdata it receives through all of its surveys and other data collection efforts as confidential, including the SOII and CFOI, and use the data only for statistical purposes. Most of the surveys BLS conducts are voluntary and the agency relies on employers and others to participate and provide accurate and complete data, and believes that keeping their information confidential is vital to maintaining respondents' trust and cooperation.

Since 2002, BLS has also been subject to the Confidential Information Protection and Statistical Efficiency Act (CIPSEA), federal legislation establishing uniform federal policy for the treatment of statistical data collected by federal statistical agencies under a pledge of confidentiality (PL 107-347). Under CIPSEA and the Office of Management and Budget (OMB) implementing guidelines (OMB, 2007), statistical agencies are required to protect data collected under a pledge of confidentiality for solely statistical purposes and ensure that data are kept confidential and not used for any nonstatistical purposes without the informed consent of the respondent.

Nonstatistical purposes include any use of the data for regulatory or enforcement purposes. BLS is permitted under CIPSEA to provide access for its "agents," including external researchers, to access confidential data; however, BLS must ensure that the data remain confidential and are only used for statistical purposes.

BLS has designated the data collected under both the SOII and CFOI as confidential and provided a pledge to respondents that all data will be used only for statistical purposes, maintained as confidential, and will not be released in identifiable form without informed consent[2] (BLS, 2017).

Unlike other BLS surveys, the SOII is mandatory and the data on which the survey is based come directly from injury and illness records that are mandated by OSHA regulations. The workplace records on which the survey is based are not considered or treated as confidential by OSHA and are available to employers, employees, former employees, unions, OSHA and other government agencies. However, when these data are collected by BLS under CIPSEA, they are confidential and BLS can permit OSHA, other agencies, and researchers to use data only for statistical purposes and enforces this through a confidentiality agreement. Therefore, OSHA is not able to use the BLS collected data for regulatory or enforcement purposes and, as discussed below and in Chapter 6, has established a separate parallel collection system to seek similar establishment-level injury and illness data from employers.

## OCCUPATIONAL SAFETY AND HEALTH ADMINISTRATION

OSHA was created by the OSH Act of 1970. Under the Act, OSHA is responsible to "assure safe and healthful working conditions for working men and women by setting and enforcing standards and by providing training, outreach, education and assistance" (OSHA, 2017a). The Act also gave states the option to set up their own regulatory enforcement programs, subject to federal OSHA approval and oversight, and currently 26 states have OSHA-approved state plans in place (OSHA, 2017b). As specified in the OSH Act, OSHA covers most private-sector employers and their employees. State and local public-sector workers are covered only if a state has an approved state OSHA program. Federal employees are covered by

[2]According to the Bureau of Labor Statistics, the CIPSEA confidentiality pledge for the SOII is as follows: "The Bureau of Labor Statistics, its employees, agents, and partner statistical agencies, will use the information you provide for statistical purposes only and will hold the information in confidence to the full extent permitted by law. In accordance with the Confidential Information Protection and Statistical Efficiency Act of 2002 (Title 5 of Public Law 107-347) and other applicable Federal laws, your responses will not be disclosed in identifiable form without your informed consent" (BLS, 2017).

federal OSHA under Executive Order 12196. Self-employed individuals are among those excluded from the Act.[3]

Data sources compiled by OSHA include those that result from worksite inspections and data submitted by employers. These data assist in evaluating the safety of a workplace, understanding industry hazards, and identifying and implementing worker protections to reduce and eliminate hazards and prevent future work-related injuries and illnesses. All employers are required to notify OSHA within 8 hours of when an employee is killed on the job or within 24 hours when an employee has an in-patient hospitalization, amputation, or loss of an eye (OSHA, 2017c).

Workplace recordkeeping requirements are in place for work-related injuries and illnesses that require more than first aid in many worksites with more than 10 employees. Covered employers are required to keep a log of all work-related injuries and illnesses (OSHA Form 300), to keep more detailed reports on individual cases (OSHA Form 301), and to prepare an annual summary of injuries and illnesses (OSHA Form 300A), which must be posted from February through April of the following year at the worksite (see Appendix E). Employers are required to maintain these records for 5 years.[4] The annual summary, OSHA 300 Log, and partial information from the OSHA 301 case reports must be made available to employees, former employees, and their representatives at the worksite if requested. OSHA and NIOSH have the right to access all of the injury and illness records that are required to be maintained.

The OSHA 300 Log and detailed OSHA 301 case reports are utilized by many employers, employees, and employee representatives for surveillance of injuries, illnesses, and hazards at the worksite. These reports also form the basis for the SOII. Employers in industries with a lower risk of serious injuries are exempt from these OSHA recordkeeping requirements, but may be required to participate in the BLS survey.[5]

---

[3]The OSH Act provides that employees excluded from OSHA's regulatory and enforcement coverage may be included in data collection and statistical programs.

[4]In December 2016, OSHA issued a regulation clarifying that employers had an obligation to ensure the accuracy and completeness of injury records during the 5-year record retention period and that this requirement was enforceable by OSHA (2016a). In March 2017, Congress overturned this regulation under the Congressional Review Act, thereby limiting OSHA's authority to enforce the obligation to keep complete and accurate records for a 6-month period after the occurrence of the injury. This action significantly limits OSHA's ability to enforce injury and illness recordkeeping requirements, and there is deep concern it will lead to less complete and accurate reporting of work-related injuries and illnesses.

[5]Industries exempted from OSHA injury and illness recordkeeping requirements are those that have a rate of injuries resulting in lost workday, restricted activity, or job transfer that is less than 75 percent of the national average, as calculated over a 3-year period. All employers, regardless of exemption for any reason, must report to OSHA any workplace incident that results in a fatality, in-patient hospitalization, amputation, or loss of an eye (29 CFR

From 1996 to 2011, the OSHA Data Initiative (ODI) collected, analyzed, and disseminated summary information on work-related illnesses and injuries from employers within specific industries and with specific-size workforces (OSHA, 2017d). These data were used to determine injury and illness rates that were establishment-specific; furthermore, when combined with other data sources these data were used to target enforcement and compliance assistance activities (OSHA, 2017i).

In 2017, a new electronic reporting system was scheduled to be implemented to provide online submission of the standard employer data on injury and illnesses under a new OSHA regulation issued in 2016 (OSHA, 2017f). This electronic injury reporting system is similar to the OSHA Data Initiative but covers a larger number of employers and requires the submission of more detailed injury and illness information from larger employers. The electronic reporting system, like the ODI, includes a number of employers that are also are part of the BLS SOII sample and required to report similar injury and illness data separately to both OSHA and BLS, since BLS cannot share the establishment specific data it collects under CIPSEA with OSHA. As will be discussed in Chapter 6, this results in duplicate reporting by a large number of employers. Accordingly, OSHA and BLS are evaluating how to best address this issue.

OSHA, through standards for toxic materials and harmful physical agents issued under Section 6(b)(5) of the OSH Act, also requires employers to conduct exposure monitoring and to provide medical surveillance of workers in jobs covered by its agent-specific standards. Currently, there are about 30 agent- or hazard-specific standards that require periodic exposure monitoring or medical examinations (OSHA, 2014). In addition, OSHA's regulation on Access to Employee Exposure and Medical Records (29 CFR 1910.1020) requires employers to maintain exposure and medical records produced under these standards as well as other exposure and medical records, and to provide these records to employees. OSHA and NIOSH also have a right of access to this information, but there is no requirement for employers to report this exposure or medical information to the agencies (see Figure 3-1).[6]

OSHA also conducts exposure monitoring as part of its health compliance inspections. The sampling results include data on personal, area, and bulk samples for a wide variety of air contaminants. OSHA has developed a Chemical Exposure Health Data website where the OSHA sampling da-

§1904.39). The list of industries exempted from routine injury recordkeeping requirements can be found online (OSHA, 2017e).

[6]Work-related illnesses that are required to be recorded under OSHA's injury and illness recordkeeping requirements will have to be reported to OSHA under the new electronic reporting requirements.

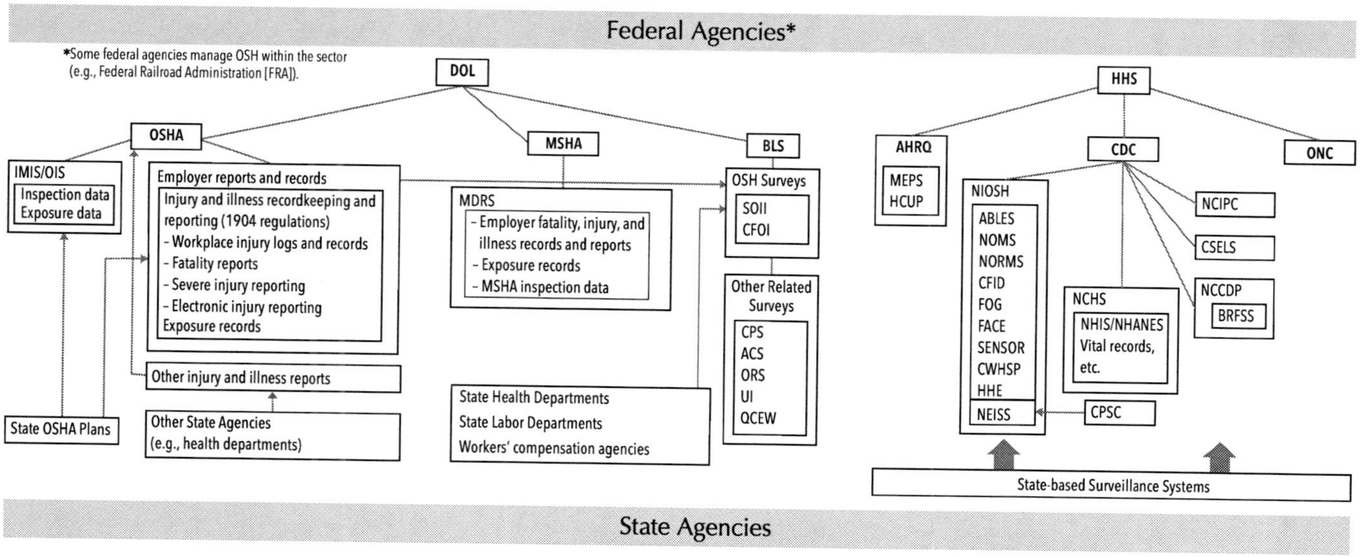


**FIGURE 3-1** Occupational Safety and Health Surveillance Agencies and Systems. The acronyms and abbreviations are listed after the table of contents of this report.

tabase from 1984 to 2015 can be searched or downloaded in its entirety (OSHA, 2017g).

## NATIONAL INSTITUTE FOR OCCUPATIONAL SAFETY AND HEALTH

Created in 1970 through the OSH Act, NIOSH was charged with carrying out the Secretary of Health and Human Services' responsibilities under that law. This agency operating under the Centers for Disease Control and Prevention (CDC) is charged in Sections 20 and 21 of the OSH Act to conduct research, experiments, and demonstrations relating to occupational safety and health; to develop criteria for recommended standards; and to conduct education programs to "provide an adequate supply of qualified personnel to carry out the purposes of this Act (Section 21)." In Section 24, the statistics section, NIOSH is given a consultative role to provide input to the Secretary of Labor to "develop and maintain an effective program of collection, compilation, and analysis of occupational safety and health statistics." NIOSH is also charged to work in cooperation with the Department of Labor on the development of injury and illnesses recording and reporting regulations.

Under the Mine Safety and Health Act, NIOSH has specific responsibilities related to mine safety and health. NIOSH is charged with conducting research to assess adverse health effects, to develop control technologies to address for mine safety and health hazards, and to make recommendations to the MSHA for improvements in mine safety and health standards. NIOSH also oversees the Coal Workers' Health Surveillance Program, which it has administered since its inception in 1970, to "prevent early coal workers' pneumoconiosis from progressing to a disabling disease. Through the program, eligible miners can obtain periodic chest radiographs (NIOSH, 2017).

NIOSH's 2016-2020 strategic goals and objectives include to "track work-related hazards, exposures, illnesses and injuries for prevention" (NIOSH, 2016). To identify research priorities NIOSH applies an approach based on assessing burden, need, and impact. Surveillance data are a key element to this process.

Since 1996, NIOSH has organized much of its work through the National Occupational Research Agenda (NORA). The first decade of NORA was organized around 21 "focus" areas organized in three categories: diseases and prevention, work environment and exposures, and research tools and approaches. In the second decade the agenda was organized on 10 industry sectors, and beginning in the third decade (2016) 7 health-based cross-sectors (e.g., hearing loss prevention and respiratory health) were added to the 10 industry sectors. Surveillance is considered as a crosscutting concern relevant to all sectors and health outcomes, and NIOSH uses an industry-health outcome matrix in surveillance planning.

Recognizing that the SOII is not a good source of information on chronic occupational diseases, the 1987 NRC report recommended that NIOSH be "the lead agency having the responsibility for the development of a comprehensive occupational disease surveillance system that would include the compilation, analysis, and dissemination of occupational illness data. . . . To accomplish this, NIOSH should request, and Congress approve, appropriation of additional funds" (NRC, 1987, p. 108). While the funding has not been made available to fulfill this recommendation, NIOSH has undertaken a number of surveillance activities to generate information not available through either the SOII or the CFOI.

NIOSH currently has a multipronged strategy to address disease and injury surveillance needs that includes

- Leveraging existing surveys and data systems managed by other agencies,
- Building occupational health surveillance capacity at the state level,
- Incorporating industry and occupation into existing surveys and other data systems,
- Improving the capacity and accuracy of autocoding tools essential to fully implement the above strategies, and
- Accelerating communication for prevention (Schnorr, 2016).

NIOSH efforts in surveillance include ongoing surveillance activities as well as surveillance research aimed at developing new tools and methods or more in-depth analyses of surveillance data (see Chapter 4). Intramural researchers apply through a competitive application process for funds to conduct surveillance research. In addition, NIOSH funds multiple states and universities to conduct surveillance and surveillance research through its extramural program.

As described in greater depth in Chapters 4 and 6, NIOSH works collaboratively with state and federal partners to implement, support, and build on existing population-based data sources, thereby strengthening the population-level data needed to provide nationwide insights on worker safety and health. Additionally, NIOSH supports a range of case-based surveillance programs focused on mortality or specific illnesses or industries that warrant more immediate action and/or follow-up to more fully characterize the problems (e.g., the Fatality Assessment and Control Evaluation Program and the Adult Blood Lead Epidemiology and Surveillance Program [ABLES]). In the past, NIOSH conducted surveys to collect data on occupational exposures to chemical, physical, and biological hazards in a representative sample of workplaces nationwide.[7] Two more recent hazard surveillance efforts focused on a web-based survey of hazards in health care settings (see Chapter 5) and collecting a limited amount of information about work organization, psychosocial, and nonspecific information about common workplace chemical and physical agent exposures and other common workplace hazards through the National Health Interview Survey (NIOSH, 2015). Efforts are under way by NIOSH and partner organizations to continue to refine and harness health-related and information technologies (e.g., electronic health records and tools for assigning standardized codes to narrative text) to improve data on work-related injuries and illnesses, to make data available to other stakeholders, and to effectively disseminate timely surveillance findings (see Chapters 4 and 6).

In addition to its core surveillance activities, NIOSH engages in other activities that can assist with surveillance efforts. NIOSH conducts health hazard evaluations which provide useful information about workplace exposures and related health impacts. NIOSH also develops and disseminates hazard alerts, sometimes jointly with OSHA, to provide information about recently identified occupational health and safety problems and exposures of particular concern, and recommend controls measures.

## MINE SAFETY AND HEALTH ADMINISTRATION

MSHA was established by the Mine Safety and Health Act (P.L. 95-164) in 1977. The agency develops and enforces safety and health rules for all mines in the United States—coal, metal, and nonmetal; underground and surface—regardless of size, number of employees, commodity mined, or method of extraction (MSHA, 2017a). MSHA regulations (30 CFR Part 50) require "mine operators to immediately notify MSHA of accidents, require operators to investigate accidents, and restrict disturbance of accident-related areas. The regulations require operators to file reports with MSHA pertaining to accidents, occupational injuries, and occupational illnesses, as well as employment and coal production data" (MSHA, 2017b). MSHA also is obligated to inspect all underground mines four times a year and all surface mines twice a year for safety and health compliance.

Coal mine operators are also required to conduct regular monitoring and reporting of coal dust exposures. MSHA maintains a comprehensive Mine Data Retrieval System accessible on the MSHA webpage that provides detailed mine-by-mine data for all mines and contractors (MSHA, 2017c). This database is also available with detailed information in a single file for health data; injury, illness, and fatality data; exposure data; results of industrial hygiene sampling data; and inspection data.

---

[7]National Occupational Hazard Survey (1972-1974), National Occupational Exposure Survey (1981-1983), and National Occupational Health Survey of Mining (1984-1989).

## OTHER FEDERAL AGENCIES

A number of other federal agencies have statutory responsibilities for occupational and public safety and health oversight for specific industries, operations, or hazards. Under section 4(b)(1) of the OSH Act, except where provided by statute or interagency agreement, OSHA does not overlap OSH responsibilities with these other agencies (Dale and Shudtz, 2013). In the transportation sector, the Federal Aviation Administration, Federal Railroad Administration, and Federal Motor Carrier Safety Administration respectively regulate and provide oversight in the aviation, rail, and trucking industries. In the maritime sector, the Coast Guard is responsible for occupational safety and health on the high seas and on the outer continental shelf, and the Department of the Interior provides safety oversight on offshore drilling platforms.

The Department of Energy (DOE) enforces safety and health at the DOE national laboratories and weapons plants and the Nuclear Regulatory Commission has responsibility for the safety of employees exposed to nuclear materials at nuclear power plants. The Environmental Protection Agency (EPA) has occupational safety and health responsibilities under the statutes it administers. The protection of farmworkers and applicators from pesticide exposures is governed by the Federal Insecticide, Fungicide and Rodenticide Act (P.L. 61-152). Under the Toxic Substances Control Act (P.L. 94-469), workers are one of the populations to be protected from the health risks from toxic chemical exposures.

In addition to providing regulatory and enforcement oversight for worker safety, many of these agencies collect data and implement surveillance systems. For example, the Federal Railroad Administration requires reporting of occupational injuries and illnesses by rail operators. EPA requires the reporting of safety and health studies, data, and notices of substantial risk, and maintains an inventory of chemicals that are manufactured in and imported into the United States.

A number of other federal agencies also collect data and conduct surveillance activities relevant and useful to OSH surveillance but are not employer or establishment based. Among these are many other centers and programs within the Centers for Disease Control and Prevention (e.g., the National Center for Health Statistics, the National Program of Cancer Registries, and the National Notifiable Diseases Surveillance System), the Agency for Healthcare Research and Quality, and the Consumer Product Safety Commission (see Box 3-1 and Chapters 4 and 6).

Given the large number of agencies, data sources, and systems, with relevant data there are great challenges in ensuring that the data are collected in a manner that allows their use for broader OSH surveillance purposes.

**BOX 3-1**
**Examples of Other Federal Agencies that Provide Data Sources for OSH Surveillance**

**Agency for Healthcare Research and Quality**

- The Medical Expenditure Panel Survey compiles annual data on how Americans use and pay for medical care. This survey has been used to examine medical care utilization and expenditures for work-related injury health conditions.
- The Healthcare Cost and Utilization Project maintains a Nationwide Emergency Department Sample and a National Inpatient Sample. Both these samples have been used for OSH surveillance.

**Centers for Disease Control and Prevention**

- The National Program of Cancer Registries collects data for the U.S. population nationwide. Data include information on cancer incidence, the type of initial treatment, and outcomes. Information about usual occupation and industry is often collected at the state level but not included in the national database.
- The Behavioral Risk Factor Surveillance System is a telephone survey system that collects information about U.S. residents and their health-related risk behaviors, chronic health conditions, and use of preventive services. The system completes more than 400,000 interviews of adults each year. The system serves as a model for similar systems in other countries. Multiple states are piloting collecting of information about current industry and occupation in the survey.
- The National Notifiable Diseases Surveillance System enables the sharing of the notifiable disease-related health information at the local, state, territorial, federal, and international government levels. The data are used to monitor, control, and prevent the occurrence and spread of infectious and noninfectious diseases and conditions. Some information about occupation and/or industry is collected for some conditions.
- The Pregnancy Risk Assessment Monitoring System collects data on attitudes and experiences of mothers before, during, and shortly after pregnancy. Data are used to identify groups of women and infants at high health risk, and monitor changes in health status. The system covers 83 percent of all U.S. births.

**Centers for Disease Control and Prevention, National Center for Health Statistics (NCHS)**

As the major statistical agency within CDC, NCHS conducts the following population-based surveys and data-collection efforts relevant to worker safety and health:

## STATE AGENCIES

State agencies have the potential to play a pivotal role in OSH surveillance. While national surveillance is essential to set national prevention and research priorities and inform federal policy development, state-specific occupational health and safety problems can be obscured in national statistics,

- The National Health Interview Survey, an annual household survey of approximately 35,000 households, was supplemented in 2010 and 2015 with a NIOSH-funded occupational health supplement.
- The National Health and Nutrition Examination Survey conducts interviews and physical examinations of approximately 5,000 individuals each year. Several supplemental modules have been developed and funded by NIOSH that focus on collecting occupational health data.

**Consumer Product Safety Commission**

- The National Electronic Injury Surveillance System is a national probability sample of hospitals in the United States that collects data from emergency room visits involving injuries from consumer products. NIOSH has funded efforts at a subset of the sample hospitals to focus on work-related nonfatal injuries.

**National Highway Traffic Safety Administration**

- The Fatality Analysis Reporting System provides data on fatal traffic crashes in the United States and includes data on work-zone crashes and types of vehicles.

**Department of Justice**

- The National Crime Victimization Survey includes information on whether injuries suffered during a crime occurred at the crime victim's workplace. The Federal Bureau of Investigation, through the Uniform Crime Reporting Program, collects and publishes the report *Law Enforcement Officers Killed and Assaulted.*

**U.S. Fire Administration**

- The National Fire Incident Reporting System collects a range of data on fires that includes firefighter injuries and fatalities as a result of the fire as well as injuries or deaths that occur at any time while on duty.

**Nuclear Regulatory Commission**

- The Radiation Exposure Information and Reporting System provides information on workforce radiation exposure at specific licensed facilities, which includes licensee employees and facility visitors.

SOURCES: NCHS, 2015, 2016; CDC, 2017a,b.

making state-based surveillance an essential supplement to national surveillance. With advances in information technology, there are an increasing number of state health data sources useful for tracking work-related injuries and illnesses that can provide both case-level and population-based information not available through systems overseen by BLS (see Box 3-2). These include, among others, hospital discharge records, laboratory reports,

**BOX 3-2**
**Examples of State Health Data Sources Used for Occupational Health and Safety Surveillance**

**Case Reports**

- Health care providers
- Hospitals and emergency departments
- Clinical laboratories
- Poison control centers

**Administrative Data**

- Hospital discharge data
- Emergency department data
- Hospital outpatient data
- Emergency medical service data
- Workers' compensation data

**State Registries**

- Birth and death data
- Cancer registries
- Birth defect registries
- Trauma registries
- Burn registries

**Surveys and Other Public Health Surveillance Systems**

- State SOII data
- Behavioral Risk Factor Surveillance System
- Youth Risk Behavior Surveillance System
- Other state-specific health surveys
- State infectious disease surveillance systems
- State syndromic surveillance systems
- State violent death reporting systems

**Other Sources**

- Media searches
- OSHA records
- Coast Guard records
- Autopsy reports

**Potential New Sources**

- All payer claims data
- Electronic health records

medical records, poison control data, and workers' compensation data. State agencies in a limited number of states have used these data sources both to generate timely, locally relevant information and to help fills gaps in surveillance at the national level. State agencies are also uniquely positioned to use the data directly to improve worker safety and health. Again in a limited number of states, states have responded to identified OSH concerns by collaborating with community partners—employers and trade associations, unions and worker centers, health care providers, and other community organizations—as well as federal and state OSHA and other government agencies to translate surveillance findings into preventive action.

OSH surveillance activity and agency leadership in OSH surveillance at the state level varies from state to state. About half of the states have state-based OSH surveillance programs with epidemiologic expertise to conduct surveillance (NIOSH, 2012). Most are based in state public health departments with a few located in labor departments or in universities that serve as the bona fide agents for state public health agencies. Only about 20 percent of states have substantial OSH surveillance capacity (CSTE, 2013). Almost all states, usually state labor departments, partner under contract with BLS to collect data for the BLS CFOI and SOII programs.

In the United States, non-work-related public health surveillance is a collaborative federal-state endeavor. State public health agencies have the legal authority to require disease reporting and forward the data to CDC with personal identifiers removed. National data on chronic and communicable disease disseminated by CDC depend on the state's legal authority and activity to collect the data. This state activity is performed with substantial federal support to all 50 states. Historically, public health surveillance was focused primarily on surveillance of communicable diseases, but has expanded in recent decades to track other health outcomes, such as cancer and violent deaths and to a more limited extent work-related injuries and illnesses. Many states have legal authority to either initiate or expand occupational disease surveillance activity (Freund et al., 1990). Some state health agencies, largely with funding from NIOSH, have built on this authority to develop case- and population-based occupational injury or illness surveillance systems and carry out related intervention and prevention activities.

As of 2017, NIOSH provides funding through a competitive funding process to 27 states, largely to public health agencies to conduct OSH surveillance and to promote the use of the data for action to address identified health and safety problems (see Figure 3-2). Twenty six states receive limited "fundamental" support for capacity building in occupational public health. At minimum, these states are working to generate the standardized occupational health indicators and build working relationships for prevention with OSH and other state public health stakeholders (Thomsen et al.,

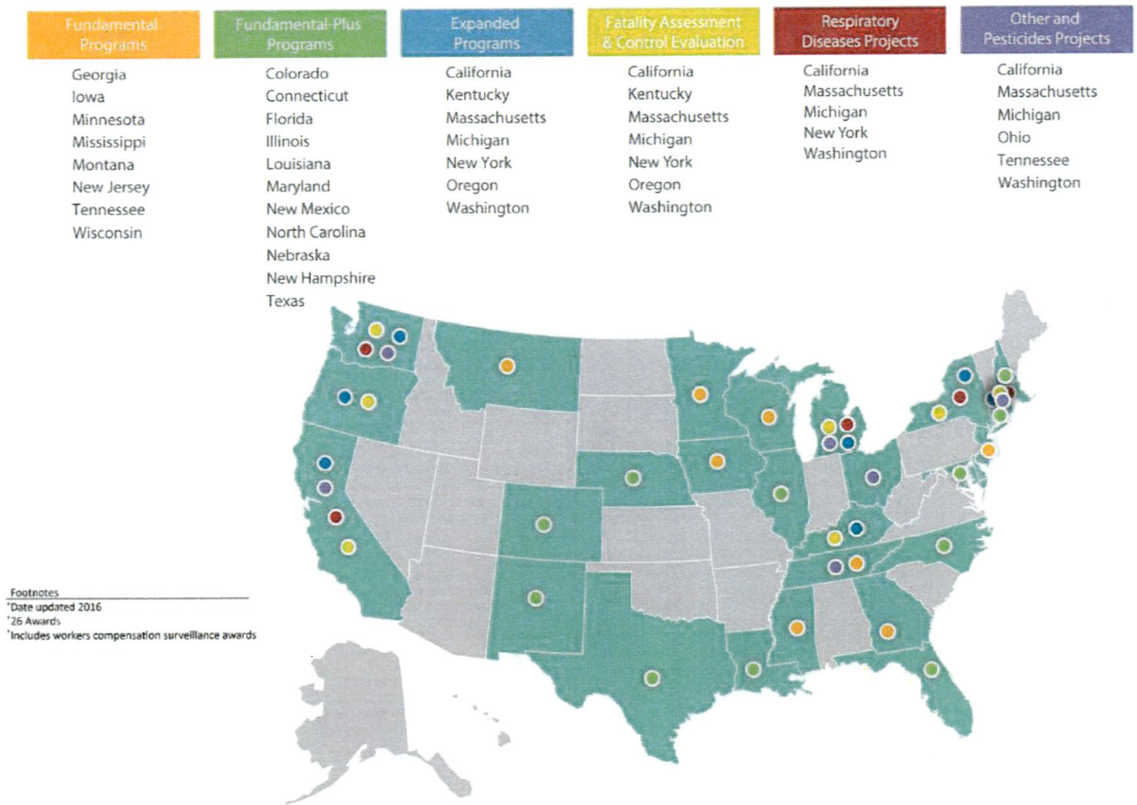


FIGURE 3-2 NIOSH-sponsored state occupational health and safety surveillance programs. SOURCE: Inserra, 2016.

2007; Stanbury et al., 2008). Seven states conduct "expanded" surveillance activities focused on targeted health outcomes, industries, or populations.[8] Some state OSH surveillance programs also partner with other public health programs to develop more comprehensive approaches to problems, such as indoor air and chemical exposures in schools, distracted driving, infectious diseases, and, more recently, opioid overdoses, that can affect workers and the general public alike (Davis and Souza, 2009). Five states are also supported by NIOSH specifically to increase the state's capacity to use workers' compensation data for prevention, providing epidemiological expertise and a focus that is frequently lacking in state agencies that administer these systems (see below and Chapter 6). State OSH programs can also serve as important bridges between labor and public health agencies within the states that have overlapping responsibilities to protect the health of the population, although the level of such interagency collaboration varies widely by state.

Workers' compensation programs are another source of state-based injury and illness information (see Chapter 4 and 6). Workers' compensation provides no-fault medical and income replacement benefits to workers injured on the job while protecting employers from liability lawsuits for such injuries. For work-related fatalities, family members may receive compensation. Most states have had workers' compensation laws in effect since the 1910s to 1920s (Sengupta and Baldwin, 2015). The National Academy of Social Insurance estimates that workers' compensation programs covered 132.7 million U.S. workers in 2013, approximately 91 percent of the 146 million civilian workers in 2014 (Baldwin and McLaren, 2016).

Data from workers' compensation cases have the potential to provide insights into trends in occupational safety and health (see Chapter 6). However, there is wide variation across the states in the nature and extent of data that are collected, including differences in the definition of reportable conditions, the type of data collected, data accessibility, and data-validation methodologies, creating significant challenges in using such data for national OSH surveillance.

## EMPLOYEES, EMPLOYERS, AND OTHER STAKEHOLDERS

While public health surveillance is primarily a function of government agencies, employers, employees, and other stakeholders also collect and use data to improve worker health and safety.

As noted above, OSHA requires companies in covered industries to maintain logs of work-related injuries and illnesses. Additionally, a number

[8] EPA contributes funding for pesticide surveillance in some states and the CFOI program is conducted by health agencies in several additional states.

of OSHA standards require employers in workplaces where regulated hazards are present to monitor exposure levels of workers. In many larger workplaces, employers implement safety and health programs, which utilize injury, illness, and exposure data to identify hazards and to take corrective action for prevention. Safety and health programs need meaningful participation of employees to be effective and are strongly recommended by federal OSHA (OSHA, 2016b) and many safety and health organizations (OSHA, 2017h). Some are required by a number of state OSHA plans, including California and Minnesota (OSHA, 2012). Recommended best practices for health and safety programs include collection, review, and use of data on work-related injuries, illnesses, and hazards for both hazard identification and performance monitoring.

To further promote and improve worker safety and health, some companies go beyond the basic regulations to focus on how best to identify potential hazards in their specific area of work and to track their safety record. Some companies conduct extensive surveillance, linking OSHA injury records with medical reports, workers' compensation reports, and other data to identify hazards and emerging problems and to track progress. For example, the Ford Motor Company, with the cooperation and involvement of the UAW (United Automobile, Aerospace and Agricultural Implement Workers of America), developed an integrated safety and health surveillance system that captures and codes every case from the medical department, OSHA logs, and incident investigations. The corporate-wide data system allows detailed analysis by plant and department, type and source of injury, and other criteria (Reeve, 2016).

In the construction industry, information on injuries, illnesses, and workers' compensation is used by contracting entities and project managers to assess the qualifications of contractors and subcontractors and to monitor project safety and health performance. Harvard University developed a database of the safety and health records of all the contractors and subcontractors working on its construction projects and a software system for tracking and analysis. The safety surveillance system now also provides services to other outside companies to help manage the safety and health oversight of their construction activities (Burke, 2016). In the commercial construction industry, many large general contractors centralize OSHA log data for all subcontractors onsite.

Despite these positive examples, it has often been challenging to engage the employer community and to encourage them to participate voluntarily in surveillance systems such as those discussed in this report. Many are reluctant to provide information to government entities given their experience with, and reservations about, the current OSH surveillance processes. The existing need to gain employer trust and engagement became clear as the committee heard testimony indicating that some employers were dissatisfied with the agencies and their current processes, questioning agency assertions that the regulatory requirements are aimed at trying to help employers to improve safety in the workplace. Some employers perceive the relationship with the agencies as adversarial rather than collaborative, characterizing the regulations as too often being coercive, focusing on enforcement and perceived "shaming" of employers rather than on developing novel methods to help employers and employees to understand the true root causes of an injury or illness. Some express concern that the agencies have based their actions on scientific literature that is biased in favor of their preferred methods and outcome while questioning the validity and credibility of employer-sponsored research.

Notwithstanding these doubts, employers welcome a smarter, better coordinated, and more cost effective set of surveillance approaches that provides useful, timely information to employers about their workplaces while recognizing and respecting current or additional privacy and confidentiality protections. A system that helps foster greater participation by the workforce, employers, and the community at large would also likely be more attractive to employers. The committee has accordingly considered these issues and offered findings, conclusions, and recommendations that are intended to support the development of optimal trust and confidence between the regulatory agencies and employers.

An example of such an approach to compliance that is supported by some employers is the Federal Aviation Administration's (FAA's) "Just Culture" approach (Reason, 1997; GAIN, 2004). In the 1990s, a "no blame culture" had developed in an effort to replace the largely punitive safety culture that it had sought to replace and it acknowledged that many unsafe acts were "honest errors." However, this no blame approach failed to address those who willfully engaged in unsafe behavior or address culpability (GAIN, 2004). The "Just Culture" notion differs from the "no blame culture" in that the former stresses that safety compliance and prevention would be based on a problem-solving approach (i.e., engagement, root-cause analysis, transparency, and information exchange) in which the primary goal is to enhance the safety performance of the individual and organization and where finding fault or penalizing noncompliance is a secondary concern. Such methods, however, do require agencies to confront individuals or organizations that willfully and sometimes repeatedly engage in practices that observers would recognize as being likely to increase the risk of a bad outcome. This also requires a clear and accepted method for distinguishing between culpable and nonculpable unsafe acts. The goal of such an approach is to create a trusted culture that encourages and even rewards people for identifying and sharing safety-related information.

Workers groups and unions are also important partners in data collection and analysis and utilization. For example, CPWR—The Center for Construction Research and Training, affiliated with North America's Building Trades Unions—conducts research, training, and service programs in the construction industry (CPWR, 2017). CPWR serves as the National Construction Research Center for NIOSH and through a NIOSH cooperative agreement conducts research to identify existing and emerging hazards and to develop evidence-based technologies and work practices to prevent injuries and illnesses. As part of its work, the center produces *The Construction Chart Book*, which presents extensive analysis of data on construction safety and health and other facets of the U.S. construction industry with creative use of publicly available information including economic, demographic, employment and income, and education and training data (CPWR, 2013).

Worker groups and unions regularly utilize injury, illness, and fatality information to seek changes in safety and health practices and stronger safety and health regulations and laws. For example, in Massachusetts, following the deaths of several Vietnamese floor refinishers, the Massachusetts Coalition for Occupational Safety and Health and the Vietnamese American Initiative for Development partnered with state and local agencies and industry groups to educate workers, employers, and consumers about the fire hazards of certain floor refinishing products and to stop the use of these products (MA COSH, 2005). These efforts ultimately resulted in new state regulations on floor refinishing product content.

## FEDERAL AND STATE AGENCY COORDINATION, COLLABORATION, INFORMATION EXCHANGE, AND PUBLIC ENGAGEMENT ON OSH SURVEILLANCE

Ideally, the OSH surveillance activities of BLS, OSHA, NIOSH, other federal agencies, and state agencies would be carried out under the framework of a national OSH surveillance strategy and through a unified surveillance system, with close coordination and collaboration on the development of strategy, planning, implementation, and evaluation. However, currently there is no overall strategy, single system, or overall formal body responsible for coordination or integration of OSH surveillance activities and programs. Just as there are a wide variety of different types of data collection and surveillance activities being conducted across agencies and organizations, there are a wide variety of mechanisms and means for coordination and collaboration among and between the agencies and from stakeholder and public input. In response to a query from the committee, BLS, NIOSH, and OSHA provided examples of coordinating activities that are summarized in the following section.

### Coordination Among BLS, OSHA, NIOSH, and Federal Agencies

BLS has some formal mechanisms for coordinating programs and work with other agencies, but according to the agency much of this work is informal. BLS provides OSHA a formal briefing the day before releasing the annual fatality and injury and illness reports. Both OSHA and NIOSH are provided a copy of the CFOI research file each year.

BLS and NIOSH have worked together informally on a number of joint surveys, such as the survey on workplace violence prevention, and research papers. For example, BLS and NIOSH staff recently collaborated with the National Highway Traffic Safety Administration (NHTSA) on a peer-reviewed article describing a methodologic approach to matching CFOI data with NHTSA's Fatality Analysis Reporting System. BLS has also provided technical assistance to NIOSH on the development of an occupational autocoding tool and has worked closely with OSHA on the development of their data-capture tools for OSHA's electronic injury reporting initiative.

OSHA and NIOSH also have both formal and informal mechanisms for collaboration and information sharing. Through the NIOSH-OSHA Liaison Information Exchange, OSHA and NIOSH staff regularly discuss topics of current concern and technical issues. OSHA and NIOSH are also members of an interagency committee that includes OSHA, MSHA, NIOSH, EPA, and the National Institute of Environmental Health Sciences that meets twice a year to share information and coordinate activities on proposed rules, risk assessments, and risk management strategies for controlling exposures to toxic agents.

OSHA, BLS, and NIOSH also have data-sharing arrangements. For example, OSHA provides the BLS CFOI program with quarterly files of fatalities investigated, and NIOSH and OSHA have an agreement under which NIOSH has access to OSHA's Information System and OSHA to ABLES data.

As discussed above, it is BLS policy to treat the microdata it collects under the SOII and CFOI as subject to the confidentiality provisions of CIPSEA. Thus this microdata are only shared with OSHA, NIOSH, and other federal agencies, subject to the same terms of confidentiality, greatly restricting and limiting the use of the BLS collected injury and illness data for surveillance, intervention, and prevention purposes.

### Coordination and Collaboration Among NIOSH and Other CDC and HHS Agencies

Underlying the accomplishments and challenges in current OSH surveillance efforts is the relationship between NIOSH and CDC. As described, NIOSH was created by the Occupational Safety and Health Act of 1970 as an agency in the Department of Health and Human Services charged with carrying out the responsibilities of the Secretary of HHS under the Act. It was administratively established as part of the CDC, the agency which has the primary responsibility for carrying out the federal government's public health functions. But the "fit" between NIOSH and CDC has always been a bit strained. The Act created NIOSH as a sister agency to OSHA, with responsibility to provide recommendations and support to the Department of Labor for carrying out regulatory activities, data collection, and statistical functions. NIOSH has often been viewed as an adjunct of OSHA and DOL, not a public health agency, and worker safety and health seen largely as a responsibility of the Department of Labor. Consequently, occupational health is not viewed by all as a "public health" issue and has not been effectively integrated into general public health either at the national or state level. Occupational health and occupational health surveillance have remained low priorities in the general public health community, including within CDC. Funding and support for occupational health surveillance has been limited. Historically, it has been difficult for NIOSH to integrate occupational safety and health into other CDC and HHS programs, and occupational health and safety has not received strong support from HHS or CDC.

In recent years NIOSH has made some strides to initiate increased collaboration and activity with other CDC and HHS agencies in a wide range of surveillance programs and activities. NIOSH is a participant in several CDC-wide surveillance groups, including CDC's Surveillance Leadership Board, CDC's Surveillance Data Platform Workgroup, which works to make essential surveillance systems more adaptable to changes in technology, knowledge, and stakeholders (including states); and CDC's Surveillance Science Advisory Group.

NIOSH also collaborates with other CDC centers to integrate occupational safety and health-related issues into broader surveillance activities. For example, NIOSH is working with the CDC's Center for Surveillance, Epidemiology, and Laboratory Services (CSELS) and the National Center for Immunization and Respiratory Diseases to include industry and occupation as a standard set of data in the National Notifiable Diseases Surveillance System. NIOSH continuously collaborates with the National Center for Health Statistics (NCHS), most recently in the inclusion of an occupational health supplement in the 2015 National Health Interview Survey and the Asthma Supplement in the Behavioral Risk Factor Surveillance System (BRFSS) designed to collect data on work-related asthma. NIOSH also collaborates closely with the CDC's National Center for Injury Prevention and Control (NCIPC), participating in biannual meetings and cross-reviewing articles of common interest prior to publication. NIOSH and NCIPC both provide funding to the Consumer Product Safety Commission's National Electronic Injury Surveillance System (NEISS), which enables both programs to obtain more complete data from NEISS than would be possible otherwise. Additionally, the NIOSH Electronic Health Record Working Group participates in national efforts by CSELS, the Office of the National Coordinator for Health Information Technology, and external partners (e.g., the Council of State and Territorial Epidemiologists [CSTE][9]) to establish end-to-end occupational case data collection from electronic health records.

### BLS, OSHA, and NIOSH Collaboration and Coordination with State Agencies

BLS, OSHA, and NIOSH have established relationships with state agencies that, among other efforts, work to carry out OSH data-collection and surveillance activities. BLS's primary collaboration is through its agreements with state partners to collect and analyze data for the SOII and CFOI. As resources permit, BLS holds a national conference with state partners and participates in annual meetings with NIOSH-funded state-based surveillance programs and CSTE.

Through OSHA's formal relationship with the 26 OSHA state plan states, the states input their inspection data into the OSHA Information System, creating a unified national database with inspection and violation data and severe injury and fatality reports. Several public health and workers' compensation agencies in individual states have developed close working relationships with state and federal OSHA programs. For example, the Michigan Department of Health and Human Services requires hospitals and emergency departments to report work-related injuries. The department reviews the medical records of the injuries and regularly makes referrals to the Michigan OSHA program to conduct follow-up inspections on workers who have amputations, burns, and skull fractures (Kica and Rosenman, 2012, 2014; Largo and Rosenman, 2015). In Massachusetts, Region I of federal OSHA, which covers the New England states, receives reports of amputations from the Massachusetts Department of Public Health, which may result in follow-up or inspections. In Region VII, the Omaha, Nebraska Area OSHA office receives a regular feed of workers' compensation cases from the Nebraska Workers' Compensation Court, which is used in targeting enforcement inspections under a local emphasis program.

[9]CSTE is a national organization of state, local, tribal, and territorial epidemiologists. It plays a key role nationally in developing national definitions for use in surveillance and determining which reportable health conditions are to be reported voluntarily to CDC. It supports effective public health surveillance and good epidemiological practice through training, capacity development, and peer consultation and by promoting collaboration between federal and state surveillance programs.

NIOSH provides both financial and technical support to state agencies for OSH surveillance activities and plays an active role in fostering coordination and collaboration (CSTE, 2011). Since 1998, NIOSH has engaged CSTE to provide a mechanism for collective state input to NIOSH's OSH surveillance planning and to facilitate surveillance capacity building and collaboration among the states, including building regional networks of states conducting surveillance. CSTE holds biannual meetings with the state-based OSH surveillance programs, including not only funded states but other states interested in building programs. CSTE also hosts ad hoc meetings bringing state OSH surveillance programs and federal agencies together to address specific topics such as collaborations with OSHA, use of BLS data systems, and closing the gaps in surveillance.

NIOSH holds periodic working meetings with states conducting expanded surveillance of NIOSH priority conditions including work-related lung disease, pesticide illness and injury, fatal injuries, and adult lead exposures. State collaboration with NIOSH intramural scientists conducting surveillance research focused on other topics is less robust. NIOSH is also working with CSTE to establish regional networks of states to build surveillance and prevention capacity in states that currently lack programs.

NIOSH has a Surveillance Coordinating Group led by the director of one of the NIOSH divisions that coordinates surveillance activities across the agency. CSTE has a state representative that participates in this group. Representatives from state programs also serve on the NIOSH Board of Scientific Counselors and NORA industry-sector councils.

In sum, there is strong collaboration between NIOSH and state-based surveillance programs that could be enhanced with additional state collaboration with NIOSH intramural researchers. State public health collaboration with the national OSHA and BLS offices has increased in recent years but remains largely informal (outside of several BLS-funded surveillance research grants to states). Collaboration with regional OSHA and BLS offices varies widely by state.

### BLS, OSHA, NIOSH, and State Agency Collaboration with Stakeholders and Public Engagement

BLS, OSHA, NIOSH, and state agencies seek to engage stakeholders and the public through a variety of ways. BLS receives regular feedback through the BLS Data Users Advisory Committee, a formal advisory committee "that provides advice to BLS from the points of view of data users from various sectors of the U.S. economy, including the labor, business, research, academic, and government communities, on matters related to the analysis, dissemination, and use of the Bureau's statistics, on its published reports, and on gaps between or the need for new Bureau statistics" (BLS,

2016b). BLS also actively seeks input from the safety and health community via the CSTE, the American Public Health Association, and others. Many of the recent changes in BLS's OSH statistics program have been driven by stakeholder input including the development of a CFOI web-scraping utility, the pilot-testing of a household survey of occupational injuries and illnesses, and the placement of SOII data in the Federal Statistical Research Data Centers (see Chapter 4).

OSHA seeks and receives stakeholder input on its major data-collection activities through the formal mechanisms established under the Paperwork Reduction Act. In addition, the collection and submission of safety and health information by employers is established by regulations, which require public notice and comment. For example, OSHA's expansion of its injury and illness reporting requirements for both severe injury reporting and electronic injury reporting were implemented through regulations for which OSHA gathered extensive input from stakeholders and the public, receiving both written comments and holding public meetings. NIOSH has formal and informal methods of obtaining stakeholder input on surveillance activities, including the NORA industry-sector and cross-sector councils, which reflect a mix of internal and external stakeholders.

Many of the state OSH surveillance programs are actively engaged with industry, labor, and other community stakeholders at the state and local levels. Some states have active program-wide advisory boards that serve as a means for obtaining ongoing stakeholder input on program activities. States also create stakeholder working groups as needed to address specific topics identified through surveillance such as injuries associated with patient handling among hospital workers in Massachusetts, injuries among workers in the logging industry in Washington State, and injuries to workers in the construction industry in California (Harrington et al., 2009; WA DLI, 2013; MA DPH, 2014).

## UPDATES ON THE RECOMMENDATIONS OF THE 1987 REPORT

The committee reviewed the 1987 NRC report, *Counting Injuries and Illnesses in the Workplace: Proposals for a Better System*, and asked the relevant federal agencies to provide an update on their progress in responding to the recommendations made in that report. In the 30 years since that report, there have been significant improvements and advances in OSH data collection and surveillance. A table outlining the recommendations and actions and developments in response to that report is included in Appendix D.

Significant developments include the Bureau of Labor Statistics' Census of Fatal Occupational Injuries, the expansion of the SOII to collect case and demographic information, and extensive research to assess and document the undercount of work-related injuries and illnesses in the SOII. OSHA instituted the OSHA Data Initiative to collect establishment-specific injury data to target inspections to the most hazardous workplaces, recently supplanted by the electronic injury reporting initiative, and has entered the exposure monitoring data collected during inspections into a publicly available database. OSHA has also implemented a new severe-injury reporting system. NIOSH has provided funding for some states to conduct surveillance of key injuries and diseases and partnered with other health agencies to conduct surveys and enhance use of existing data sources to gather information on occupational injuries and health conditions.

However, on several of the recommendations made in the 1987 report, there has been little or no progress. The recommendation for BLS to provide regular feedback to employers on the results of the BLS survey in order to benchmark performance and the recommendation to OSHA to require employer reporting of exposure monitoring data for specific substances have not been implemented. The development of a comprehensive occupational disease surveillance system with NIOSH as the lead agency has not been pursued as funding has not been available. Furthermore, OSH surveillance in most states remains at the capacity building level given the lack of recognition of OSH surveillance as a core public heath function and the attendant lack of resources.

Overall, the state of OSH surveillance has improved since the 1987 report, but significant gaps and barriers to achieving comprehensive OSH surveillance remain.

## SUMMARY AND CONCLUSIONS

Occupational safety and health surveillance in the United States is a collaborative, decentralized effort carried out by a large number of federal and state agencies with the substantial involvement of BLS, OSHA, NIOSH, MSHA, and state health and labor departments. A broad range of stakeholders including employers, employees, and safety and health professionals also participate. The data and information that are collected and utilized come from many different sources, ranging from workplace-based reporting by employers, to individual case reports by physicians, to population-based health surveys.

OSH surveillance represents a prime example of the difficulties of generating and using data for science, policy, and public information that is posed by our decentralized federal statistical system.

**Conclusion: While there is some coordination and collaboration among and between different agencies engaged in OSH surveillance, much of the agency interaction is limited to information exchange rather than joint collaborative programs and initiatives.** Collecting SOII and CFOI data under

CIPSEA greatly limits the sharing and use of these data for surveillance purposes by other federal and state agencies. It would be worthwhile for BLS, OSHA, and NIOSH to jointly explore if there are alternative arrangements for collecting and sharing establishment specific injury and illness data to make the data more widely available and useful for surveillance and prevention purposes and to also avoid duplicate reporting. An example of one such arrangement could be for OSHA to collect the data from employers and provide the data to BLS for statistical analysis and to NIOSH for research and surveillance purposes.

**Conclusion: While significant improvements in OSH surveillance have occurred since the 1987 report, significant gaps remain as OSH surveillance in the United States remains highly fragmented with no overall unified surveillance strategy or mechanism for planning, coordinating, or executing programs.** In the chapters that follow, a detailed review of the current systems, existing gaps and barriers, promising developments, and recommendations for a 21st-century OSH surveillance system are presented.

## REFERENCES

Baldwin, M. L., and C. McLaren. 2016. *Workers' Compensation: Benefits, Coverage, and Costs, 2014*. Washington, DC: National Academy of Social Insurance. Available online at https://www.nasi.org/sites/default/files/research/NASI_Workers_Comp_Report_2016.pdf (accessed June 23, 2017).

BLS (Bureau of Labor Statistics). 2015. A look at violence in the workplace against psychiatric aides and psychiatric technicians. *Monthly Labor Review*. Available online at https://www.bls.gov/opub/mlr/2015/article/pdf/a-look-at-violence-in-the-workplace-against-psychiatric-aides-and-psychiatric-technicians.pdf (accessed November 8, 2017).

BLS. 2016a. Injuries from falls to lower levels, 2013. *Monthly Labor Review*. Available online at https://www.bls.gov/opub/mlr/2016/article/pdf/injuries-from-falls-to-lower-levels-2013.pdf (accessed November 8, 2017).

BLS. 2016b. *Data users advisory committee charter*. Available online at https://www.bls.gov/advisory/duaccharter.htm (accessed June 23, 2017).

BLS. 2017. *Confidentiality of data collected by BLS for statistical purposes*. Available online at https://www.bls.gov/bls/confidentiality.htm (accessed November 8, 2017).

Burke, G. 2016. *Contractor performance measurement*. Presentation to the National Academies Committee on Developing a Smarter National Surveillance System for Occupational Safety and Health in the 21st Century, September 21. Available online at https://vimeo.com/187718133 (accessed July 10, 2017).

CDC (Centers for Disease Control and Prevention). 2017a. *Behavioral Risk Factor Surveillance System*. Available online at https://www.cdc.gov/brfss/index.html (accessed June 23, 2017).

CDC. 2017b. *What is PRAMS?* Available online at https://www.cdc.gov/prams/index.htm (accessed June 23, 2017).

CPWR (The Center for Construction Research and Training). 2013. *The Construction Chart Book, 5th Ed*. Available online at http://www.cpwr.com/publications/construction-chart-book (accessed March 25, 2017).

CPWR. 2017. About CPWR. Available online at http://www.cpwr.com/about/about-cpwr (accessed March 25, 2017).

CSTE (Council of State and Territorial Epidemiologists). 2011. *Guidance: public health referrals to OSHA*. Available online at http://www.cste2.org/webpdfs/occupational/OSHAreferral922011.pdf (accessed May 26, 2017).

CSTE. 2013. *National assessment of epidemiology capacity: Findings and recommendations*. Available online at http://www.cste2.org/2013eca/CSTEEpidemiologyCapacityAssessment2014-final2.pdf (accessed September 4, 2017).

Dale, G. N., and P. M. Shudtz, eds. 2013. *Occupational Safety and Health Law*, 3rd Ed. Arlington, VA: American Bar Association and Bloomberg BNA.

Davis, L., and K. Souza. 2009. Integrating occupational health with mainstream public health in Massachusetts: An approach to intervention. *Public Health Reports* 124(Suppl 1):5-14.

Drudi, D. 2015. The quest for meaningful and accurate occupational health and safety statistics. *Monthly Labor Review.* December 2015. Available online at https://www.bls.gov/opub/mlr/2015/article/the-quest-for-meaningful-and-accurate-occupational-health-and-safety-statistics.htm (accessed January 19, 2017).

Freund, E., P. J. Seligman, T. L. Chorba, S. K. Safford, J. G. Drachman, and H. F. Hull. 1990. Mandatory reporting of occupational diseases by clinicians. *Morbidity and Mortality Weekly Report* 39(RR-9):19-28.

GAIN (Global Aviation Information Network). 2004. *A Roadmap to a Just Culture: Enhancing the Safety Environment.* Flight Operations/ATC Operations Safety Information Sharing Working Group, Global Aviation Information Network (GAIN). Available online at https://flightsafety.org/files/just_culture.pdf (accessed January 19, 2017).

Harrington, D., B. Materna, J. Vannoy, and P. Scholz. 2009. Conducting effective tailgate trainings. *Health Promotion Practice* 10(3):359-363.

Inserra, S. 2016. Presentation at the NIOSH-State Partners Meeting, December 7, 2016, Atlanta, GA.

Kica, J., and K. D. Rosenman. 2012. Multi-source surveillance system for work-related burns. *Journal of Occupational and Environmental Medicine* 54(5):642-647.

Kica, J., and K. D. Rosenman. 2014. Multi-source surveillance system for work-related skull fractures in Michigan. *Journal of Safety Research* 51:49-56.

Largo, T. W., and K. D. Rosenman. 2015. Surveillance of work-related amputations in Michigan using multiple data sources: Results for 2006-2012. *Occupational and Environmental Medicine* 72:171-176.

MA COSH (Massachusetts Coalition for Occupational Safety and Health). 2005. *Protecting workers and homeowners from wood floor-finishing hazards.* Available online at http://www.masscosh.org/files/ProtectingFromFloorFinishingHazards.pdf (accessed March 25, 2017).

MA DPH (Massachusetts Department of Public Health). 2014. *Moving into the future: Promoting safe patient handling for worker and patient safety in Massachusetts hospitals.* Available online at http://www.mass.gov/eohhs/docs/dph/occupational-health/ergo-sph-hospitals-2014.pdf (accessed June 19, 2017).

MSHA (Mine Safety and Health Administration). 2017a. *Mission.* Available online at https://www.msha.gov/about/mission (accessed March 20, 2017).

MSHA 2017b. *Part 50 reports: Mining industry accident, injury, illness, employment, and coal production reports.* Available online at https://www.msha.gov/data-reports/reports (accessed March 30, 2017)

MSHA. 2017c. *Mine data retrieval system.* Available online at https://arlweb.msha.gov/drs/drshome.htm (accessed March 30, 2017).

NCHS (National Center for Health Statistics). 2015. *About the National Health and Nutrition Examination Survey*. Available online at https://www.cdc.gov/nchs/nhanes/about_nhanes.htm (accessed June 23, 2017).

NCHS. 2016. *About the National Health Interview Survey*. Available online at https://www.cdc.gov/nchs/nhis/about_nhis.htm (accessed June 23, 2017).

NIOSH (National Institute for Occupational Safety and Health). 2012. *Surveillance activities: NIOSH-funded research grants*. https://www.cdc.gov/niosh/programs/surv/grants.html (accessed November 29, 2017).

NIOSH. 2015. *National Health Interview Survey: Occupational Health Supplement*. Available online at https://www.cdc.gov/niosh/topics/nhis/summary.html (accessed June 19, 2017).

NIOSH. 2016. *About NIOSH*. Available online at https://www.cdc.gov/niosh/about (accessed January 22, 2017).

NIOSH. 2017. *Coal workers' health surveillance program*. Available online at https://www.cdc.gov/niosh/topics/cwhsp/ (accessed June 4, 2017).

NRC (National Research Council). 1987. *Counting Illnesses and Injuries in the Workplace: Proposals for a Better System*. Washington, DC: National Academy Press.

OMB (Office of Management and Budget). 2007. Implementation Guidance for Title V of the E-Government Act, Confidential Information Protection and Statistical Efficiency Act of 2002 (CIPSEA). Notice of Decision. 72 *Federal Register* 33362-33377.

OSHA (Occupational Safety and Health Administration). 2012. *Injury and illness prevention programs white paper.* Available online at https://www.osha.gov/dsg/topics/safetyhealth/OSHAwhite-paper-january2012sm.pdf (accessed March 25, 2017).

OSHA. 2014. *Medical screening and surveillance requirements in OSHA standards: A guide*. Available online at https://www.osha.gov/Publications/osha3162.pdf (accessed March 13, 2017).

OSHA. 2016a. Clarification of employer's continuing obligation to make and maintain an accurate record of each recordable injury. Final rule. 81 *Federal Register* 91792-91810.

OSHA. 2016b. *Recommended Practices for Safety and Health Programs*. Available online at https://www.osha.gov/shpguidelines/docs/OSHA_SHP_Recommended_Practices.pdf (accessed March 25, 2017).

OSHA. 2017a. *About OSHA*. Available online at https://www.osha.gov/about.html (accessed January 22, 2017).

OSHA. 2017b. *State plans*. Available online at https://www.osha.gov/dcsp/osp/ (accessed January 27, 2017).

OSHA. 2017c. *Report a fatality or severe injury*. Available online at https://www.osha.gov/report.html (accessed January 22, 2017).

OSHA. 2017d. *Establishment specific injury & illness data (OSHA Data Initiative)*. Available online at https://www.osha.gov/pls/odi/establishment_search.html (accessed January 23, 2017).

OSHA. 2017e. *Non-mandatory appendix A to subpart B — Partially exempt industries*. Available online at https://www.osha.gov/recordkeeping/ppt1/RK1exempttable.html (accessed August 21, 2017).

OSHA. 2017f. *Final rule issued to improve tracking of workplace injuries and illnesses*. Available online at https://www.osha.gov/recordkeeping/finalrule/index.html (accessed January 23, 2017).

OSHA. 2017g. *Chemical exposure health data*. Available online at https://www.osha.gov/opengov/healthsamples.html (accessed March 25, 2017).

OSHA. 2017h. *Safety and health program voluntary standards*. Available online at https://www.osha.gov/shpcampaign/voluntary-standards.html (accessed March 25, 2017).

OSHA. 2017i. *OSHA Injury and illness recordkeeping and reporting requirements*. Available online at https://www.osha.gov/recordkeeping/index.html (accessed December 20, 2017).

Reason, J. T. 1997. *Managing the Risks of Organizational Accidents*. Farnham, UK: Ashgate.

Reeve, G. R. 2016. *Occupational health surveillance in manufacturing*. Presentation to the National Academies Committee on Developing a Smarter National Surveillance System for Occupational Safety and Health in the 21st Century, September 21. Available online at https://vimeo.com/187718132 (accessed July 10, 2017).

Schnorr, T. 2016. *NIOSH surveillance program: Overview*. Presentation to the National Academies Committee on Developing a Smarter National Surveillance System for Occupational Safety and Health in the 21st Century, June 15. Available online at https://vimeo.com/172464267 (accessed July 10, 2017).

Sengupta, I., and M. L. Baldwin. 2015. *Workers' Compensation: Benefits, Coverage, and Costs, 2013*. Washington, DC: National Academy of Social Insurance. Available online at https://www.nasi.org/research/2015/report-workers-compensation-benefits-coverage-costs-2013 (accessed March 25, 2017).

Stanbury, M. J., H. Anderson, P. Rogers, D. Bonauto, L. Davis, B. Materna, and K. D. Rosenman. 2008. *Guidelines for Minimum and Comprehensive State-based Public Health Activities in Occupational Safety and Health 2008*. DHHS (NIOSH) Publication No. 2008-148. Available online at https://www.cdc.gov/niosh/docs/2008-148/pdfs/2008-148.pdf (accessed June 23, 2017).

Thomsen, C., J. McClain, K. Rosenman, and L. Davis. 2007. Indicators for occupational health surveillance. *Morbidity and Mortality Weekly Report* 56(RR-1):1-7.

WA DLI (Washington State Department of Labor and Industries). 2013. *ESSB 5744: Report on the Development and Implementation of the Logger Safety Initiative—2013 Report to the Legislature*. Available online at http://www.lni.wa.gov/Main/LoggerSafety/pdfs/LoggerSafetyInitiative.pdf (accessed June 19, 2017).

# 4

# Current Status of Federal and State Programs and Cross-cutting Issues

## INTRODUCTION

A wide range of health outcomes—from fatal falls and amputations to chronic lung disease, musculoskeletal disorders, and cancer—are either caused or exacerbated by hazardous exposures in the work environment. Currently, there is no single, comprehensive occupational safety and health (OSH) surveillance system in the United States but rather an evolving set of systems using a variety of data sources that meet different surveillance objectives, each with strengths and weaknesses. As discussed, many federal and state agencies carry out this work. Figure 4-1 shows a Venn diagram representing major data sources for occupational injuries and illnesses (under the major categories of employers, medical records, and individuals) and the overlap among the systems that collect these data for occupational injury and illness surveillance. The degree of overlap is a best guess estimate of how much overlap there is between systems and illustrates that in the potential smart surveillance system of the future, there will be some overlap between the sources and that there will be a need to collect data from multiple sources to obtain a comprehensive picture of OSH problems. In Figure 4-1, the sources of data from employers are the Bureau of Labor Statistics' Survey of Occupational Injuries and Illnesses (BLS SOII), workers' compensation, and direct reports by employers to regulatory agencies. The sources of data from medical records are ambulance companies, audiometry providers, birth certificates, cancer registries, death certificates/medical examiners, hospitals; emergency departments, clinics, clinicians, laboratories, Medicare databases, and poison control centers. The sources

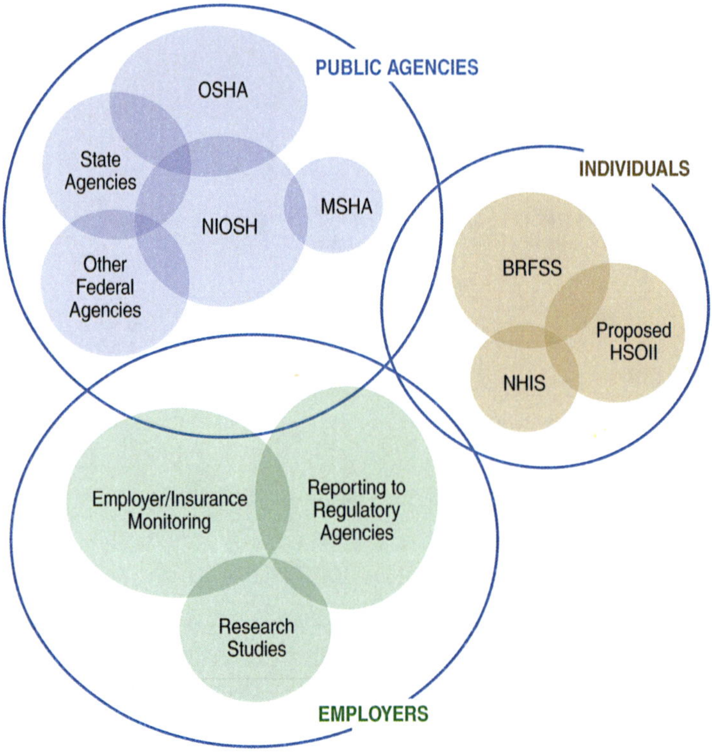


**FIGURE 4-1** Current Possible Overlaps in Data Sharing from the Major Sources of Data Used for Occupational Injury and Illness Surveillance. The size of inner circles do not represent relative importance of the source. The Census of Fatal Occupational Injuries is the only system to use data from multiple sources, and is illustrated by the overlaps. Other sources may overlap, but it is uncertain whether there is overlap because of confidentiality.

of data from individuals (current workers, retirees, disabled) are health surveys (BRFSS, National Health Interview Survey [NHIS], National Health and Nutrition Examination Survey [NHANES]), and proposed Household Survey of Nonfatal Occupational Injuries and Illnesses (HSOII). The other sources of data are newspaper reports/electronic media. There are limited data to show overlap between different sources, and examples of studies showing the degree of overlap are: BLS SOII and workers' compensation (Rosenman et al., 2006; Boden and Ozonoff, 2008); or individual conditions such as acute traumatic fatalities (BLS, 2016a), amputations (Largo and Rosenman, 2015), burns (Kica and Rosenman, 2012) or skull fractures (Kica and Rosenman, 2014). The successful model for a multisource occupational surveillance system is the Census for Fatal Occupational Injuries (CFOI).

The major systems in place are largely focused on injury and disease outcomes. Hazard and exposure surveillance is important but currently very limited. In this chapter, we provide an overview of the major systems, organized by the health outcome under surveillance: fatal injuries, nonfatal injuries, and diseases. The current status of hazard surveillance is also discussed. The chapter concludes with a discussion of several crosscutting issues in OSH surveillance. A number of promising new developments in the field are described in Chapter 6.

## SURVEILLANCE OF FATAL OCCUPATIONAL INJURIES

Substantial advances in the surveillance of fatal occupational injuries have been made over the past several decades with the development and implementation of several surveillance systems and programs (see Table 4-1).

In 1992, BLS (in the U.S. Department of Labor [DOL])—in response to recommendations in the 1987 National Research Council (NRC) report—established the Census of Fatal Occupational Injuries (CFOI), a nationwide surveillance system designed to produce a timely census of all fatal work injuries in the United States.[1] A federal-state cooperative program implemented in all 50 states, CFOI uses multiple data sources, such as death certificates, police reports, federal agency administrative data, workers' compensation claim records, and news media, to identify, verify, and describe fatal work injuries. For each death, information is collected about the deceased worker, including occupation and demographic characteristics, the establishment, the equipment involved, and the circumstances of the event. Two or more independent source documents are used to confirm that fatal injury cases are work related. National and state-level findings including both counts and rates[2] by industry and occupation and other worker and establishment characteristics are issued annually, in the following calendar year. Surveillance findings are published through the media and made available on an interactive BLS website (BLS, 2017a). BLS has also published more extensive analyses of fatalities among workers in specific industries (e.g., road construction and landscaping), among specific worker groups (e.g., Hispanics, Asians), and due to specific events (e.g., machine-related deaths and suicides), as well as studies comparing analytic methods

[1]BLS notes that "to be included in CFOI: a death must have resulted from a traumatic injury; the incident that led to the death must have occurred in the United States, its territories, or its territorial waters or airspace; and it must be related to work. Defining work-relatedness is complex and BLS CFOI applies a standard definition" (BLS, 2016a).

[2]Rates exclude persons younger than 16 years of age, volunteers, and resident military personnel.

**TABLE 4-1** Examples of Surveillance Systems for Fatal Occupational Injuries in the United States

| Surveillance System | Scope[a] | Type[b] | Responsible Agency(ies) | Condition(s) under Surveillance[c] |
|---|---|---|---|---|
| Census of Fatal Occupational Injuries (CFOI) | National<br>50 states | P | BLS*<br>States* | O: Fatal occupational injuries |
| Fatality Assessment and Control Evaluation (FACE) Program | 7 states | C | NIOSH*<br>States | O: Fatal occupational injuries—targeted incidents |
| Firefighter Fatality Investigation and Prevention Program | National<br>State | C | NIOSH* | O: Fatal occupational injuries and heart attacks in line of duty |
| Commercial Fishing Incident Database | National<br>Regional<br>State | P | NIOSH* | O: Fatal occupational injuries and incident survivors |
| Fatalities in Oil and Gas Extraction | National | C | NIOSH* | O: Fatal occupational injuries and illnesses |

NOTES: BLS, Bureau of Labor Statistics; NIOSH, National Institute for Occupational Safety and Health; OSHA, Occupational Safety and Health Administration; asterisk denotes funding agency.

[a]Geographic levels at which findings are publicly available.

[b]P, population based: data are collected on a census or representative sample of a defined population and allow for assessing extent of a health related event and monitoring trends with

(Windau, 1998; Sincavage, 2005; Wiatrowski, 2005; Byler, 2013; Pegula, 2013; Harris, 2016). The National Institute for Occupational Safety and Health (NIOSH) also has access to the CFOI microlevel files to conduct more in-depth analysis and respond to public information requests, but limited resources to do this work. Given that BLS collects CFOI under the Confidential Information Protection and Statistical Efficiency Act (CIPSEA), CFOI data can only be used for statistical purposes (OMB, 2007).

BLS continues to make enhancements to CFOI and has, for example,

| Data Source(s) | Population Covered[d] | Approach | Time Frame for Report Release |
|---|---|---|---|
| Multiple: OSHA, Coast Guard records, death certificates, police reports, media, etc. | All workers | Census | Annual |
| Field investigations and other data sources | All workers | Case series (~2% all worker deaths in U.S.) | Ongoing |
| Field investigations and other data sources | All firefighters | Case series (~40% all firefighter deaths in U.S.) | Ongoing |
| Multiple: Coast Guard records, death certificates, local media, etc. | All workers in commercial fishing industry | Census | Intermittent |
| Multiple: OSHA records, motor vehicle crash reports, police reports, media and professional contacts | All oil and gas extraction workers | Census | Intermittent |

this population over time/locale; C, case based: focus of data collection is on individual cases that require follow-up or immediate public health action. These approaches are not mutually exclusive.

[c]O, outcome; H, hazard; E, exposure.

[d]The population covered may include active and former workers, retirees, and others depending on the system.

expanded the number of variables collected to include birthplace, contracted worker status, and, in the case of a contracted decedent, ownership of the contracting firm. Distinguishing fatal work-related motor vehicle incidents from non-work-related incidents is particularly difficult, and BLS is currently working with NIOSH, the National Highway Traffic Safety Administration (NHTSA) and state agency partners to pilot new approaches to provide more comprehensive information about fatal occupational crashes.

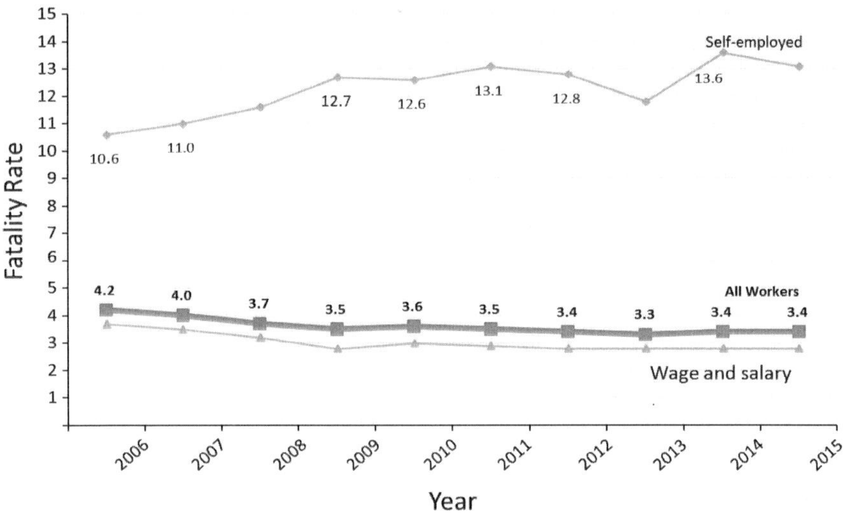


FIGURE 4-2 Rate of fatal work injuries per 100 full-time equivalents by employee status, 2006-2015. The 2015 rate of fatal work injuries for all workers was 3.4 fatal work injuries per 100,000 full-time equivalent workers (FTEs). The rate for self-employed workers has consistently been higher than that of all workers since the adoption of hours-based rates. NOTE: Rate = (Fatal work injuries/Total hours worked by all workers) × 200,000,000, where 200,000,000 = base for 100,000 FTEs working 40 hours per week, 50 weeks per year. The total-hours-worked figures are annual average estimates of total at work multiplied by average hours for civilians, 16 years of age and over, from the Current Population Survey (CPS). In 2008, CFOI implemented a new methodology, using hours worked for fatal work injury rate calculations rather than employment. For additional information on the fatal work injury rate methodology, please see BLS, 2010. SOURCE: BLS, 2016b.

BLS has also improved the timeliness of releasing CFOI findings to the public.

CFOI is well respected as an authoritative count of fatal work injuries at the national and state levels and a model for multisource surveillance of a health outcome. It provides robust information about the burden and distribution of fatal occupational injuries over time and serves as an important example of how surveillance findings can be used to monitor progress in meeting prevention goals, target intervention activities, and set research priorities. Findings have identified a decline in the fatal occupational injury rate over time as well as continuing high risk among workers in certain industries, such as fishing, construction, transportation, and the self-employed (BLS, 2016b) (Figure 4-2). The data have also brought to light high-risk populations, including older workers and Hispanic workers (Byler, 2013; BLS, 2016b). Findings have also led to new federal outreach initiatives including, for example, a nationwide campaign to prevent falls in construction (OSHA, 2017a) and increased Spanish-language assistance for Hispanic employers and workers (OSHA, 2007).

While CFOI provides essential statistical data on the approximately 5,000 fatal occupational injuries that occur each year (BLS, 2017a), the CFOI data lack sufficient detail about underlying causes of fatal incidents needed to develop specific prevention recommendations, and, as described above, BLS confidentiality practices restrict the use of the data for case-level public health intervention. For instance, the system does not allow for case-based follow-up to intervene in specific workplaces to protect others at risk or to learn more about specific factors (e.g., names of specific chemicals) contributing to workplace injuries.

Since 1991, NIOSH has supported a collaborative endeavor with the states, the Fatality Assessment and Control Evaluation (FACE) program, to conduct in-depth investigations of targeted fatal occupational incidents with the objective of identifying factors contributing to these deaths. Currently seven states are funded by NIOSH to conduct approximately 100 investigations each year.[3]

NIOSH identifies national targets for investigation, which currently include falls from elevations and machine-related deaths, and the participating states have the option of identifying their own targets. FACE investigators use an approach developed by Haddon to identify underlying causes of the incidents (Haddon, 1970). This information is used to develop comprehensive recommendations for prevention. Each investigation results in a report with an incident description and prevention recommendations which is disseminated widely to industry, labor, equipment manufacturers, and other stakeholders. State FACE programs also work with local agency and private-sector partners to promote implementation of recommendations (NIOSH, 2017a). FACE provides valuable in-depth information about the circumstances leading to deaths that is either not collected or because of confidentiality not available in the CFOI statistics and exemplifies how data can be used to inform prevention. The investigation of sentinel fatalities through the FACE program has helped identify and increase public awareness of previously unrecognized hazards, and led to improvements in OSH practices at worksites, changes in public policy, and development of new, safer technologies. For example, FACE investigations in Michigan led to a nationwide alert on hazards of methylene chloride exposures associated

[3]Some additional state health agencies, such as in Wyoming, that do not participate in the FACE program also track work-related deaths for public health intervention purposes (WY DWS, 2016).

with bathtub refinishing (MI FACE, 2013; OSHA, 2013); FACE investigations of deaths among floor finishers in Massachusetts contributed to state law banning the use of highly flammable floor sealing products (MA COSH 2005; Azaroff et al., 2011); and a Kentucky FACE investigation of the death of an auto technician led to the redesign of a handicapped-accessible accelerator pedal (CSTE, 2015a).

In 1998, Congress funded NIOSH to establish a Firefighter Fatality and Investigation Program modeled on FACE in which NIOSH staff conducts in-depth investigations of select firefighter deaths throughout the country (NIOSH, 2017b). Targets for this program go beyond fatal injuries to include outcomes such as heart attacks in the line of duty. NIOSH also supports more in-depth data collection on fatal incidents in several high-risk industries, including fishing and oil and gas extraction, and works with partners in these targeted industries to disseminate findings and promote use of the data for prevention (NIOSH, 2012, 2014a, 2017c).

In recent years, there has been increased interest by state health agencies, unions, community organizations, and others for more timely and detailed information on specific workplace fatalities. In response, the Occupational Safety and Health Administration (OSHA) made additional timely information (including the victim's name) on workplace fatalities investigated by the agency[4] available on its website until changing this practice mid-2017 (OSHA, 2017b). Inability to access CFOI data for purposes of public health intervention has been a challenge. BLS has taken some steps to address this and make available to the public and state public health agencies timely data on fatal occupational injuries that are available through public sources including the web (Pegula and Measure, 2016). Several nonprofit organizations have made data on work-related fatalities available through interactive mapping applications (CPWR, 2017; National COSH, 2017).

## SURVEILLANCE OF NONFATAL OCCUPATIONAL INJURIES

Nonfatal occupational injuries as discussed in this chapter encompass traumatic injuries due to sudden events such as falls, motor vehicle crashes, violence, and being struck by machinery. Also included within this discussion are musculoskeletal disorders (MSDs), which are defined by NIOSH as "soft-tissue injuries that may be caused by either sudden or sustained

[4]Only about a third of all fatal occupational injuries are investigated by OSHA (AFL-CIO, 2017; MA DPH, 2017); the remaining deaths are either outside of OSHA's jurisdiction (e.g., self-employed, public sector in federal OSHA states), are due to causes such as on-the-road motor vehicle deaths or homicides at work not routinely investigated by the agency, or came to OSHA's attention more than 6 months after the incident.

exposure to repetitive motion, force, vibration, or awkward positions"[5] (NIOSH, 2016g).

Over the past several decades, federal agencies, in collaboration with states, have made improvements in surveillance of nonfatal occupational injuries. Major activities include the following:

- BLS enhancements to the SOII to collect additional data on more severe injuries, add data on public employees, and expand data analysis;
- OSHA's severe injury reporting system;
- NIOSH's leveraging of other national public health surveillance systems and surveys; and
- State-based projects, funded largely by NIOSH, using state data sources to conduct surveillance of nonfatal occupational injuries at the state level.

## Survey of Occupational Injuries and Illnesses

The SOII is the only source of uniform nationwide statistics on nonfatal work-related injuries in the United States, providing annual estimates of the number and rates of work-related injuries and illnesses for the nation and for the states that collaborate with BLS. This is available by industry, establishment employment size, and case severity as measured by work restriction or days of work lost (BLS, 2016c) (Figure 4-3). The SOII is conducted by BLS in collaboration with state agencies in 41 states, territories (Guam, Puerto Rico, and the U.S. Virgin Islands), and the District of Columbia, and data is collected directly by BLS in the other 9 nonparticipating states. The SOII is an annual nationwide survey with a sample of approximately 250,000 private industry and state and local government establishments, selected to represent all industries and all sizes of establishments. Data

[5]The classification of MSDs in different surveillance systems is complicated. While in both the International Classification of Disease and the Occupational Injury and Illness Classification system used by BLS most MSDs are classified as injuries, there is a small set of conditions generally included in a broad definition of MSDs, such as carpal tunnel syndrome and Raynaud's syndrome, that are classified as diseases. Since 2011, BLS and OSHA have applied a surveillance case definition for MSDs in employer-reported data aimed at identifying cases due to sustained exposure that takes into account information about both the nature of injury or illness and the event or exposure. Included are cases "where the nature of the injury or illness is pinched nerve; herniated disc; meniscus tear; sprains, strains, tears; hernia (traumatic and nontraumatic); pain, swelling, and numbness; carpal or tarsal tunnel syndrome; Raynaud's syndrome or phenomenon; musculoskeletal system and connective tissue diseases and disorders, when the event or exposure leading to the injury or illness is overexertion and bodily reaction, unspecified; overexertion involving outside sources; repetitive motion involving micro tasks; other and multiple exertions or bodily reactions; and rubbed, abraded, or jarred by vibration" (NIOSH, 2016a).

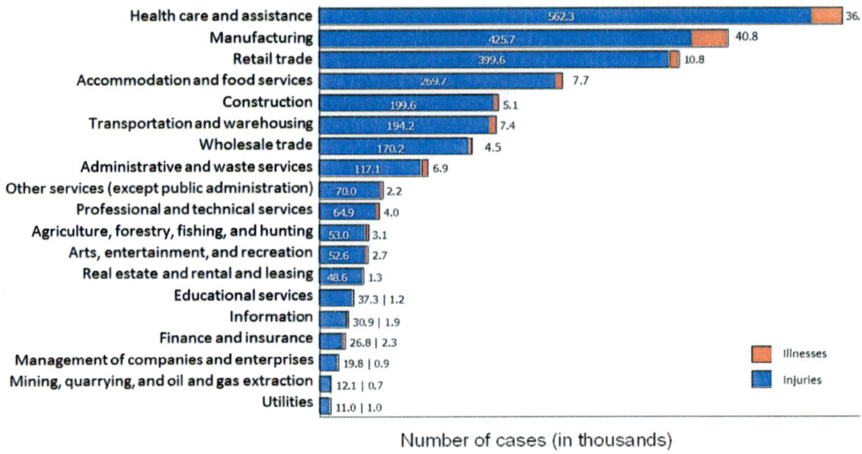


**FIGURE 4-3** Distribution of nonfatal occupational injuries and illnesses by private industry sector, 2015. Injuries accounted for the majority of cases reported in 2015 among individual private industry sectors. Illnesses accounted for only a small fraction of cases reported in each industry sector. SOURCE: BLS, 2016d.

from the railroad and mining industries included in the SOII published findings are obtained by BLS from the Federal Railroad Administration and the Mine Safety and Health Administration, which require reporting of all fatal and nonfatal work-related injuries and illnesses by railroad and mine employers.

The data collected by BLS for the SOII are based upon OSHA required records for occupational injuries and illnesses—the OSHA Log of Work-Related Injuries and Illnesses (Form 300), the Injury and Illness Incident Report (Form 301), and the Summary of Work-Related Injuries and Illnesses (Form 300A). Establishments selected for the SOII are required by law to maintain records of all work-related injuries and illness that meet the OSHA requirements for recordkeeping for 1 year (even if otherwise excluded from OSHA injury recordkeeping requirements) and report that information to BLS. Injuries and illnesses that are recordable for OSHA include those resulting in loss of consciousness, medical treatment beyond first aid, one or more missed days from work, restricted work activities, or transfer to another job (MI DCH, 2013). Since 1992, in addition to information from the OSHA Log, as recommended in the 1987 NRC report, the SOII has collected data on the nature and circumstances of the injury or illness (OSHA Form 301), as well as the characteristics of the affected workers for injuries resulting in one or more days away from work, which currently account for 32 percent of all reported cases (BLS, 2016e). Given the changes in case management that can shift injured workers from lost-workday cases to restricted-work and job transfer cases, it is challenging to use these statistics as measures of injury severity.

This more detailed information is referred to as the *case and demographic data*. BLS collects data elements that employers are required by OSHA to record, along with several optional case and demographic variables including the race and ethnicity of the injured worker and the category that best describes the regular type of job or work. These variables, however, are generally not analyzed due to the limited reporting of these optional data (Wiatrowski, 2014). As with CFOI, national and state-level estimates are issued annually, and are published and made available on the web (BLS, 2016c).

BLS has continued to make improvements in the SOII. For example, in 2006, BLS began generating and publishing experimental rates by occupation, age, and gender, in addition to routinely published rates by industry, establishment, and case severity.[6] In addition, starting in 2008, BLS expanded the scope of the SOII by collecting and reporting data about work-related injuries and illnesses among state and local government workers, who were previously not included in the nationwide survey.[7] BLS has also piloted the collection of more detailed case and demographic data on a sample of injuries and illnesses that did not result in lost time but led to job transfer or restriction—which account for an increasing proportion of all cases reported over time—21 percent in 2015 (Wiatrowski, 2014; BLS, 2016f). Attention to enhanced information on all types of injuries is planned to be sought through a proposed household survey (see Chapter 6). BLS has also updated its injury and illness classification system to reflect current workplace hazards and has substantially reduced processing time to make the data more rapidly publicly available (Wiatrowski, 2014).

The information derived from the analysis of SOII data is used by federal and state government agencies to set standards, target enforcement compliance assistance activities, develop and update educational programs, and set research priorities. Industry and unions use the data for

---

[6]Information on the number of employee hours provided by participating employers is used as the denominator in calculating rates by industry, establishment employment size, and case type. Since 2006, BLS has used external sources of denominator data (the BLS Current Population Survey and the Occupational Employment Statistics Program) to generate and publish experimental rates by occupation, age, and gender—as the employment data provided by participating employers are not broken down by these data elements.

[7]Reporting by states and local agencies is voluntary for the 22 states without federally approved OSHA state plans covering state and local government workers. While the overall survey response rate for private-sector establishments is about 95 percent, it is around 80 percent for state and local governments (Wiatrowski, 2014).

benchmarking their own injury and illness experience. In some industries, most notably construction, industry-specific rates are used as criteria in assessing qualifications of contractors and subcontractors.

The SOII, however, has a number of significant limitations. Excluded from the SOII are the self-employed (e.g., independent contractors, including gig economy workers [on-demand contractors and freelance workers]), household workers, federal workers, U.S. Postal Service workers, and workers on farms with fewer than 11 employees. Altogether, these excluded populations represent about 9 percent of the workforce, the majority of whom are self-employed (Simpson, 2016). Also, the SOII does not collect case and demographic information for 70 percent of all reported injuries and illnesses (BLS, 2016e).

While it has long been recognized that the SOII does not adequately capture chronic occupational illnesses for reasons discussed under disease surveillance below, there is increasing evidence that undercounting of injuries and acute illnesses is also a significant limitation, and there may be differential rates of reporting by establishment, injury, and incident characteristics, as well as by state. In studies to quantify undercounting on OSHA Logs, estimates of the undercount range from 20 to 70 percent (Rosenman et al., 2006; Boden and Ozonoff, 2008; Wuellner et al., 2017). There is a consensus that the SOII substantially underestimates the true burden of work-related injuries among workers employed in establishments covered by OSHA and that multiple factors contribute to underreporting (Azaroff et al., 2002; Ruser, 2008; Spieler and Wagner, 2014; Wiatrowski, 2014; Wuellner and Bonauto, 2014; Rappin et al., 2016; Wuellner and Phipps, 2016; Wuellner et al., 2016, 2017; Fagan and Hodgson, 2017) (see Box 4-1).

Given concerns about the completeness of the injury and illness counts in the SOII, in recent years BLS has supported a program of research to better understand both the extent of the undercount and the contributing factors. Individual states have also compared findings from state systems combining data from multiple data sources (i.e., multisource surveillance systems) with SOII estimates (Kica and Rosenman, 2014; Largo and Rosenman, 2015). Recently reported findings based on interviewers with a sample of recordkeepers at establishments participating in the SOII indicated that many recordkeepers possess a limited understanding of the recordkeeping requirements and identified many common recording errors. Recordkeeper characteristics (SOII experience, OSHA recordkeeping experience, and OSHA recordkeeping training) were found to be associated with better practices and knowledge (Wuellner and Phipps, 2016).

Additionally, there is concern that some of the most vulnerable workers, which often include racial and ethnic minority and immigrant workers, may be those whose injuries are least likely to be captured in the SOII

**BOX 4-1**
**Underreporting of Injuries in the Workplace**

Multiple factors have been identified as contributing to the underreporting of work-related injuries in the workplace and consequently the SOII:

- Unawareness or confusion by recordkeepers about the OSHA recordkeeping requirements.
- Lack of effective systems in workplaces for documenting and recording injuries.
- Delays between initial injury and subsequent work absence that goes into the following calendar year.
- Potentially intentional nonreporting due to concerns about OSHA penalties, increases in workers' compensation premiums, or failure to be considered for contracts due to poor safety records.
- Failure by workers to report their injuries to their employers or to file claims under the workers' compensation system, particularly if the injury is less serious. Workers may fear reprisal by their employers (2016 OSHA regulations stipulate that employer policies for reporting workplace injuries and illnesses need to be reasonable and specifically prohibit retaliation against employees who report a workplace injury).
- Employer medical management policies that may contribute to undertreating and underrecording.
- Failure of health care provider to recognize work-relatedness (especially true for musculoskeletal disorders and chronic illnesses).

(Sabbath et al., 2017). Notably, because race and ethnicity are optional variables, these data elements are often incomplete. From 2011-2016, race and ethnicity information was not reported in the SOII for 36-39 percent of the cases involving days away from work (BLS, 2018). A related challenge is the inability to characterize the injury experience of temporary agency workers, as under OSHA recordkeeping rules, most injuries to temporary agency workers are recorded under the supervisory employer, which is more commonly the host employer rather than the "staffing agency," which is the employer of record (OSHA, 2014). OSHA records and in turn the SOII are also unable to provide information across multiple employers who may be working at a single site with shared work environment responsibilities. This is an increasing concern as multiemployer workplaces, already common in the construction industry, are becoming more common in other industries (Weil, 2014, 2017).

In the last major revision to OSHA's injury and illness recordkeeping regulations in 2001, the agency had proposed that construction employers be required to maintain a site log for larger construction projects of the injuries and illnesses of all subcontractors with 11 or more employees working at the site. The proposal was widely supported by unions, but opposed by employers, and thus was not included as a provision, citing complexities for construction employers and questioning the utility of the data (OSHA, 2001).

Similarly, neither OSHA injury and illness recordkeeping and reporting requirements nor the SOII capture the enterprise-wide injury and illness experience for employers who operate multiple establishments. Thus, there is no ability to track injuries and illnesses at the enterprise or corporate level. OSHA had considered requiring enterprise level reporting by some larger employers as part of the new electronic injury reporting requirements issued in 2016 (discussed in Chapters 3 and 6), but the agency decided not to include such a provision in the final rule, again citing complexities involved with such reporting (OSHA, 2016a).

Another significant limitation of the SOII is the relatively small sample size at the state level. In all but the largest states, the sample is too small to provide sufficiently detailed data on different injury types in specific industries or occupations needed to target tailored intervention and prevention efforts (Davis et al., 2012). The SOII data are not routinely aggregated over years, although BLS is currently exploring methods for doing so. (On the other hand, in doing so trends in time would be less evident.) Finally, SOII, like CFOI, is a population-based surveillance system. BLS is required by CIPSEA to protect the confidentiality of SOII data and to ensure they are used only for statistical purposes. Establishment-level data are not made available to OSHA or other agencies for purposes of follow-up in specific workplaces to protect others at risk or for further information gathering. The SOII data are available to researchers for BLS-approved projects. In the past, this research had to be carried out at BLS headquarters but the Census Bureau and the Bureau of Labor Statistics are in the process of making certain BLS restricted data sets, including SOII and CFOI data, available to qualified researchers for statistical research exclusively through the Federal Statistical Research Data Centers managed by the U.S. Census Bureau.[8]

### *Potential Improvements in the SOII*

BLS's pilot collection of case and demographic data on the approximately 21 percent of recordable cases resulting in job transfer or restriction has demonstrated this is feasible (BLS, 2015, 2017b). Routine collection of these data in the SOII would provide important information about how

---

[8]The SOII data set is now available at the Federal Statistical Research Data Centers and the CFOI data are planned to be made available at these centers in fall 2017.

occupational injuries and illnesses are managed and a much more complete accounting of the full range of injuries and illnesses and the circumstances in which they occur (NRC, 2001). Because, this would involve utilization of data already recorded by employers under the OHSA record-keeping rules and the large majority of employers report their data to SOII electronically, it would add little to the employer reporting burden. New natural language processing tools are being developed by BLS for auto-coding narrative information collected in the SOII (see Chapter 7), which can be anticipated to continue to improve over time and to help minimize the additional costs of processing these data at BLS.

Eliminating health disparities is a major goal of public health (CDC, 2016). Collection of information about race and ethnicity in public health surveillance systems is of high importance as it provides information to identify disparities in health across population groups. While there are robust data on the differential risk of fatal occupational injury across racial and ethnic groups available through CFOI, there is a paucity of such information on nonfatal injuries and illnesses. Collection of race and ethnicity data as an optional variable in the SOII for approximately 40 percent of cases suggests it is feasible for employers to report this information. However, incomplete reporting of these data underscore the importance of OSHA's making this a mandatory data element in OSHA record-keeping. As described elsewhere, the changing nature of employment arrangements is a 21st century reality and occupational health surveillance systems will need to address this change. The current inability of the SOII to characterize the OSH risks faced by workers employed in nontraditional employment arrangements is a significant gap that can be addressed in part by collecting this information on OSHA logs and the SOII. It may benefit employers as well as provide information needed for accurate calculation of their establishment-specific illness and injury rates. Engagement of employers and workers in developing guidance on how to collect these data elements will be important.

An unheralded advantage of the SOII is its potential to foster use of data by employers and workers at the establishment level. Providing feedback to those who report data for surveillance purposes is a core component of an effective public health surveillance system (Thacker et al., 2012). Advances in information technology provide new opportunities for BLS to provide employers with feedback on their data and data analysis tools that they can use to target efforts to prevent work-related injuries and illness among their employees and reduce associated economic costs. As described, private sector employer participation in the SOII is legally mandatory. According to a 2011 analysis, the response rates of private sector employers were over 90 percent between 2003-2010 (Huband and Bobbit, 2013). While response rates for the SOII leave room for improvement, they are certainly of a lesser concern than underreporting of injuries and illnesses. Making the data meaningful for employers so that it is seen as useful rather than as simply a reporting exercise has potential to improve not only response rate but data quality. Periodic assessment of undercount will continue to be an important activity of BLS to improve the accuracy and representativeness of the SOII estimates. OSHA enforcement of injury and illness recordkeeping requirements, including enforcement to protect workers who are retaliated against for reporting injuries, will remain critical to help ensure more complete and accurate recording and reporting.

**Conclusion: Although limited, the SOII remains the most extensive system for standardized information on nonfatal occupational injuries and acute illnesses across the nation and the SOII continues to be an important source of information today.** Better and more efficient use of the SOII to meet surveillance objectives, including characterization of disparities in risk among vulnerable groups of workers, is possible.

**Recommendation A: BLS and OSHA should collaborate to enhance injury and illness recording and the SOII to achieve more complete, accurate, and robust information on the extent, distribution, and characteristics of work-related injuries and illnesses and affected workers for use at the worksite and at national and state levels.** As part of this effort, BLS should routinely collect detailed case and demographic data for injuries and illnesses resulting in job transfer or restricted-duty work. Furthermore, OSHA should amend its injury and illness recording requirements to collect information on race and ethnicity as well as on employment arrangement to identify vulnerable worker populations and risks that may be associated with the changing nature of work.

In the near term:

- OSHA should make type of employment arrangement (e.g., traditional, independent contractor, temporary agency worker, and on-call worker) and race and ethnicity mandatory data elements on the OSHA Form 301, and BLS should incorporate this information into the SOII case and demographic data. OSHA should collaborate with BLS in determining the best approach to collecting this information (e.g., what questions should be included on Form 301).
- BLS should routinely collect detailed case and demographic data for injuries and illnesses resulting in job transfer or restricted duty as well as those resulting in days away from work.
- BLS should implement methods to aggregate SOII data over time to generate more robust and detailed state-level estimates.
- OSHA and BLS should collaborate to enhance recordkeeping train-

ing for employers and BLS should evaluate approaches for providing initial information and ongoing feedback to data recorders in establishments enlisted to participate in the SOII both to improve the data quality and to promote employer use of data for prevention.

In the longer term:

- BLS and OSHA should collaborate to determine the best way to collect injury and illness data across multiple employers working at single sites, and across enterprises with multiple establishments.
- BLS should assess the feasibility and usefulness of extending collection of case and demographic data to all reported cases as automated approaches to coding SOII narratives are improved. Options for collecting such data should be evaluated in light of information that will be made available through the OSHA electronic reporting initiative.

### OSHA Severe Injury Reporting

OSHA has undertaken many initiatives over the years to address the lack of access to establishment-specific injury and illness data in the SOII. Since OSHA was established in 1971, employers under OSHA jurisdiction have been required to promptly report workplace fatalities and incidents resulting in hospitalizations of three or more employees to the agency for evaluation and investigation. In 2014, OSHA expanded its reporting rules to require employers to report fatalities within 8 hours and all incidents resulting in in-patient hospitalization, amputation, or loss of an eye within 24 hours (referred to by OSHA as *severe injuries*). These expanded reporting requirements have enabled OSHA to better target limited enforcement and compliance assistance resources to the most dangerous workplaces and engage more high-hazard employers in eliminating serious hazards (Michaels, 2016). The data also have allowed the agency to gather information on the causes of incidents to identify serious hazards, prevent future incidents, and form the basis for revised standards.

The severe injury reporting rule went into effect January 1, 2015. During its first year of operation, OSHA reported that employers notified the agency of 10,388 nonfatal incidents, including 7,636 hospitalizations and 2,644 amputations. These data represent only reports made to federal OSHA, not the OSHA state plans.[9] In addition, OSHA estimates that these

[9]The 26 OSHA state plan states are required to adopt reporting requirements that are as effective as federal OSHA's requirements. A number of states, including California and Washington, have required the reporting of all in-patient hospitalizations for many years. Other state plan states are in the process of adopting requirements similar to federal OSHA.

reports may represent fewer than half of the severe injuries that were required to have been reported to federal OSHA (Michaels, 2016). Recent findings from Massachusetts, a federal OSHA state, indicate that fewer than half of work-related amputations were reported by employers covered by OSHA (Grattan et al, 2017).

Since OSHA does not have the resources to conduct an onsite investigation or inspection for every severe injury report, the agency has developed a triage system for evaluating and responding to reports (OSHA, 2016b). Under its current procedures, all fatalities and reports of two or more hospitalizations are subject to an inspection as are reports involving a worker under 18 years of age, reports from employers with a history of similar incidents or multiple violations, or reports of incidents involving hazards covered by emphasis programs or any imminent danger. For other reports, OSHA gathers additional information from the employer about the incident, the injured employee, hazards involved, and history of the employer to determine whether to conduct an onsite inspection or an offsite rapid response investigation. Under a rapid response investigation, an employer is required to conduct its own investigation, abate any hazards found, report findings and abatement verification to OSHA in writing, and post the information in the workplace. To assist employers, OSHA is providing tools on conducting incident investigations, developed jointly with the National Safety Council.

OSHA is also collecting and compiling data from the severe injury reports. During the severe injury report intake process, the agency collects data on the employer, the employee (including employee's age and employment arrangement [e.g., contract or temporary worker]), the injury, and the incident, and then enters it into the OSHA Information System (the agency's primary program and regulatory database). Employer reports of fatalities and severe nonfatal injuries stripped of personal identifiers are made available by OSHA in a publicly accessible database (OSHA, 2017c). OSHA's summary analysis of severe injury reports is conducted on an ad hoc basis.

From these reports, OSHA has identified industries with a higher number of severe injuries (e.g., construction, support activities for mining, and drilling and servicing for oil and gas). The reports have also helped OSHA identify patterns of severe injuries and take preventive action. For example, in 2015, OSHA's Atlanta regional office noticed numerous reports of fingertip amputations among supermarket and restaurant workers using food slicers (Michaels, 2016). The agency took action to contact food service employers across the region and provide information about the hazards of food slicers and control measures to keep workers safe (OSHA, 2015).

### NIOSH Use of Existing Surveillance Systems and Surveys

NIOSH conducts a number of surveillance activities aimed at filling gaps in information on nonfatal injuries. These include efforts using existing data sources in partnership with other federal agencies as well as support for injury surveillance in states, which will be discussed in the following section.

Since the early 1990s, NIOSH has had an interagency agreement with the Consumer Product Safety Commission to collect data on nonfatal occupational injuries through an occupational supplement to the National Electronic Injury Surveillance System (NEISS). NEISS collects data on injuries associated with consumer products or work that are treated in emergency departments from a sample of U.S. hospitals. From this sample, the total number of occupational injuries treated in hospital emergency rooms nationwide can be estimated. NIOSH publishes periodic reports based on these data and makes findings from 1998 forward publicly available through an interactive web-based application (CDC, 2017a) (Box 4-2). The data are also available to researchers for approved research projects. A significant advantage of this system, referred to as "NEISS-Work," is that it captures information on all injured civilian workers who seek treatment in emergency departments, regardless of size of employer and nature of work or employment arrangement, and therefore includes workers who are excluded from the SOII, such as the self-employed or volunteers. Another advantage is that it does not require an employer to report or even be aware of the injury or for an employee to file a workers' compensation claim. Also, information is abstracted from medical records and thus can be assumed to be more clinically accurate than employer-reported data. It has been estimated that approximately one-third of work-related injuries are treated in emergency departments (Jackson, 2001). While not all work-related injuries are captured in NEISS-Work, this system provides useful national estimates of the trends in nonfatal occupational injuries overall and by nature of injury and event, and demographic characteristics (see, for example, Table 4-2). Notably, findings from NEISS-Work provide a different picture of the trend in occupational injuries than that provided by the SOII. Whereas the SOII data indicate a downward trend in occupational injury rates since the early 1990s, NEISS rates have only more recently declined (Jackson, 2001; A. Reichard, NIOSH, personal communication, 2017). SOII and NEISS-Work injury rates by age also differ markedly. The findings from these two systems differ due to differences in the scope of the population covered and reporting by employers (Marsh et al., 2016; Rosenman, 2016; Tonozzi et al., 2016). A disadvantage of NEISS-Work is that information on occupation and industry has to be collected from the often incomplete medical record, making it less useful for targeting

**BOX 4-2**
**Searchable Work-Related Injury and Illness Data Systems and Resources Maintained By NIOSH**

According to the Centers for Disease Control and Prevention (CDC), the **Work-Related Injury Statistics Query System (Work-RISQS)** "is an interactive query tool to obtain estimates for the number of nonfatal occupational injuries treated in U.S. hospital emergency departments (EDs). The data are derived from the National Electronic Injury Surveillance System—Occupational Supplement (NEISS-Work). Work-RISQS users may interactively query on worker demographic characteristics, nature of injury, and incident circumstances for ED-treated injuries from 1998 through the present to obtain national estimates" (CDC, 2017a).

The CDC states that the "**Employed Labor Force query system** provides employed worker population estimates (workers >= 15 years) from 1980 through the present. Estimates may be derived for numerous demographics and work-related characteristics including industry sectors being targeted through the NIOSH National Occupational Research Agenda (NORA). Estimates are based on a subset of the BLS CPS public access data files maintained by the NIOSH Division of Safety Research (DSR), Surveillance and Field Investigation Branch, Injury Surveillance Team" (CDC, 2017b).

Furthermore, CDC notes that "BLS developed the **Occupational Injury and Illness Classification System (OIICS)** to characterize occupational injury and illness incidents. OIICS was originally released in 1992. BLS redesigned OIICS in 2010 with subsequent revisions in 2012. The OIICS includes four hierarchical coding structures: nature of the injury or illness; part of body affected by the injury or illness; source and secondary source of the injury or illness; and event or exposure. NIOSH in collaboration with BLS has developed this website and the accompanying downloadable software application as a resource for occupational safety and health researchers, policy makers, employers, and others who may need to use the OIICS for uniformly characterizing occupational injuries and illnesses or better understanding the national occupational injury and illness data

prevention measures. Other disadvantages of NEISS-Work include a small sample size, lack of information about lost work time (other than limited information if the individual is hospitalized), and limited information about race and ethnicity. NEISS-Work does provide an important mechanism to conduct follow-back surveys of workers to obtain additional information, and NIOSH has conducted such surveys when additional resources have been available (Marsh et al., 2016). Recent follow-back studies by NIOSH found that the employers were aware of the injuries among their employees who were treated in the emergency departments and identified in the released by the BLS and NIOSH (CDC, 2017c). To code or use the OIICS coded data, one would need to properly understand the OIICS Coding Selection Rules" (BLS, 2012;CDC, 2017d).

According to NIOSH, the "**Coal Workers' Health Surveillance Program (CWHSP) Data Query System** is a federally mandated worker medical monitoring program for underground coal miners. Its intent is to prevent early coal workers' pneumoconiosis (CWP) from progressing to disabling disease. Eligible miners can obtain periodic chest radiographs. Miners found to have radiographic evidence of CWP are advised of this and are provided, by law, the opportunity to work in a "low-dust" occupation in the mine. The program is operated by NIOSH, which has maintained data from this program since its inception in 1970. The data can be queried to produce tables and maps using the interactive system. The x rays included in this system encompass x rays from the Coal Workers' X-ray Surveillance Program (CWXSP) and the Enhanced Coal Workers' Health Surveillance Program. The CWXSP also includes x rays from the National Coal Study (also known as the National Study of Coal Workers' Pneumoconiosis) and the Miners' Choice Health Screening Program" (NIOSH, 2016b).

**National Occupational Mortality Surveillance (NOMS) System**. NIOSH maintains a database (NOMS) that includes deaths by cause for most years between 1985 and 2010 where occupation and industry has been coded on death certificates from a convenience sample of states. The data are provided in a queryable form that prepares estimates of proportional mortality by industry and occupation to allow exploration of known or potential associations that could lead to research to test hypothesized associations and to examine impact of prevention activities.

**The Work-Related Lung Disease Surveillance System (eWoRLD)**. This interactive website presents data on select work-related respiratory diseases. For many of these conditions, related exposure data are also presented. Data may presented in table, chart, or map format. Information is also available on tobacco use and smoking status by industry and occupation from the National Health Information Survey.

NEISS system. Accordingly, any difference between the increased number of injuries identified in NEISS, as compared to the SOII, was not secondary to employers being unaware of the injury (Bhandari et al., 2016; Marsh et al., 2016; Tonozzi et al., 2016). NHTSA previously funded NIOSH to conduct targeted surveillance, including a follow-back survey of emergency medical service workers using NEISS-Work.

NIOSH supports additional focused activities to track nonfatal injuries in several industries. For example, as described, NIOSH funds the Center for Construction Research and Training, whose in-house data center uses a

**TABLE 4-2** Number, Percentage, and Rate of Nonfatal Occupational Ladder Fall Injuries, by Selected Characteristics and Data Source, United States, 2011

| | NEISS-Work (nonfatal injuries treated in EDs) | | | | |
|---|---|---|---|---|---|
| Characteristic | Number | (95% CI) | %[a] | Rate[b] | (95% CI) |
| **Total** | **34,000** | **(±6,800)** | **100** | **2.6** | **(±0.5)** |
| Men | 30,100 | (±6,300) | 89 | 4 | (±0.2) |
| Women | 3,900 | (±1,300) | 11 | 0.7 | (±0.2) |
| 20–34 | 11,000 | (±2,500) | 32 | 2.7 | (±0.5) |
| 35–44 | 9,900 | (±2,500) | 29 | 3.3 | (±0.6) |
| 45–54 | 7,100 | (±2,500) | 21 | 2.2 | (±0.5) |
| 55–64 | 4,400 | (±1,500) | 13 | 2.1 | (±0.5) |
| ≥65 | — | | — | — | |
| White, non-Hispanic | 19,900 | (±6,100) | 59 | 2.2 | (±0.4) |
| Other, non-Hispanic | 2,000 | (±1,000) | 6 | 0.9 | (±0.5) |
| Hispanic | 5,800 | (±2,800) | 17 | 3.1 | (±1.3) |
| Unknown | — | | — | — | |

NOTE: CI, confidence interval; ED, emergency department.
[a]Percentages might not sum to 100 because of exclusions and rounding.
[b]Per 10,000 FTE workers. Each injury is only counted once, regardless of the number of ED visits. Rates were calculated by CDC based on the number of injuries and the number of primary employed FTE workers from the BLS Current Population Survey, 2011. Variances for NEISS-Work data and CPS data were pooled to estimate the variance for injury rates.
SOURCE: Socias et al., 2014.

wide range of health, employment, and economic data sources to provide information about health and safety in the construction industry. Findings are routinely updated and published in an online chart book widely used by industry stakeholders (CPWR, 2013). Given the increasing recognition of the contribution of workplace violence to the burden of occupational injuries, NIOSH is also working with the Bureau of Justice, which conducts the National Crime Victimization Survey to improve and report on data on work-related violence.

The agricultural sector poses unique challenges for surveillance given the range of work settings from large industrial farms relying on a migrant workforce to small family farms where family members are also at risk. Additionally, the agricultural industry is exempted from many labor protection laws in the United States. Data on all farm-related fatalities are collected in the multisource CFOI, which shows agriculture to have one of highest fatality rates of any industry sector. Data on nonfatal injures among agriculture workers are collected in the SOII, however, as described, this survey excludes all farms with fewer than 11 employees. Also, the injury and illness experience of owner-operator farms and family members who work on these farms is not captured as these farms do not meet the BLS criteria for an employer. Despite limitations, the SOII indicates that agriculture has one of the highest rates of nonfatal injuries and illnesses.

In the past, to provide information not available through the SOII, NIOSH conducted surveillance of injuries and illnesses among agricultural workers through a series of surveys carried out in collaboration with U.S. Departments of Agriculture and Labor, which were discontinued in 2015 due to fiscal constraints. NIOSH is currently considering alternative approaches to conducting surveillance of health and safety of production agriculture workers, including the potential role that its extramural program, specifically, the NIOSH-funded regional Agricultural Centers, might play (NIOSH, 2016c). The new proposed BLS household survey will also need to be evaluated on its ability to identify agriculture-related injuries and illnesses. Legislative changes that allowed BLS to survey employers with less than 11 employees would be useful but only partially effective to identify nonfatal cases missed in its employer survey, since it would not address the issue of the owner-operated farm that relies on family members.

**Conclusion: Agriculture is a high-risk industry sector where a legislative restriction and particular work arrangements reduce the ability to obtain accurate counts of nonfatal injuries and illnesses.** Future evaluations will need to be conducted to determine if recommended changes are sufficient to address the current limitations in OSH surveillance in agriculture.

**BOX 4-3**
**Data to Action**

***Preventing Injuries in Michigan:*** The Michigan OSH surveillance program uses hospital data and workers' compensation data to track amputations, burns, crushing injuries, acute hospitalized injuries, farm-related injuries, and skull fractures among workers. For example, a crushing injury occurred when a male worker in his late forties had his gloved hand caught in a hydraulic power press at a furniture manufacturer. The state surveillance program received the report and the incident was referred to Michigan OSHA. The company was inspected by Michigan OSHA about 3 months after the injury occurred and had not corrected the hazard at the time of the inspection. The company was cited for two serious

*continued*

**BOX 4-3** Continued

violations (not utilizing any point-of-operation guard or device on the press and not establishing a die-setting procedure to ensure guards were properly installed and effective before the press was released for operation).

***Protecting Teens at Work:*** The Massachusetts Department of Public Health Young Workers Project (YWP) has used similar data to track injuries to working teens—a population at high risk of being injured on the job. When the surveillance data revealed high numbers of burns in a large franchised restaurant chain, interviews with injured teens and follow-up worksite investigations by the YWP found teens were injured while brewing coffee. This primarily occurred during peak hours with high demand. Not aware brewing was ongoing, teens were pulling out brew baskets, causing hot coffee slurry to splash over their hands and wrists causing second and third degree burns. The YWP recommended that the company work with its equipment suppliers to find an engineering solution to the problem. The company designed a brewer which locks the coffee funnel until brewing is complete which is being used in many of the franchised establishments across the country. Other YWP findings have been used to promote increased protections for youths under Massachusetts' child labor laws, requirements that youth job training programs provide health and safety training, and statewide outreach to teens, parents, teachers, and employers about protecting youth at work.

***Industry-Based Surveillance in the Trucking Industry—Washington State's TIRES Project:*** In Washington State, the trucking industry has high workers' compensation claim rates and costs. Past research by Washington's state-based OSH surveillance program indicated that the most common and costly injuries were musculoskeletal disorders and falls. In response to this problem, Washington developed the Trucking Injury Reduction Emphasis through Surveillance (TIRES) project. TIRES partnered with trucking industry stakeholders including employer associations and unions to further define the causes of work-related injuries and develop educational materials. Case-based surveillance that included interviews with injured workers led to the identification of loading and unloading freight, decoupling trailers, strapping down loads, and ingress and egress from the cab and trailer as foci for injury prevention. With the steering committee's oversight and input, TIRES developed resources for prevention and education. The products focus on identifying workplace hazards and providing low-cost, simple solutions for injury prevention. During fiscal year 2016, more than 150,000 people accessed TIRES publications online via the project website (NIOSH, 2015a). TIRES training materials are used as a major component of the Washington Teamsters Training programs. TIRES researchers also developed an online simulation training for young truck drivers, safety directors, and company leaders, in response to requests for more interactive education. These tools have been downloaded more than 30,000 times all over the world, including by TIRES partners in the Alabama Trucking Association Workers' Compensation Fund, the Motor Carriers of Montana, and the Safety Driven-Trucking Safety Council of British Columbia (NIOSH, 2015a).

### State-Based Surveillance of Nonfatal Occupational Injuries

As described in Chapter 3, a small number of states, have established expanded case- and population-based surveillance systems for select occupational injuries and illnesses. While most of these expanded state programs focus on occupational illnesses, several states conduct injury surveillance and prevention activities addressing targeted injury types, populations at risk, or industries. State-based work-related injury surveillance makes use of records from a wide array of data sources—hospitals, emergency departments, poison control centers, and workers' compensation systems—to identify and track injuries and to target intervention and prevention activities (see examples in Box 4-3).

Several of these focused state systems combine data to get a more comprehensive picture of the condition under surveillance than would be possible with any single data source, and all include intervention and prevention activities. Half of the fundamental state surveillance programs, supported by NIOSH, use available state data sources to generate a standard set of over 20 occupational health indicators (Thomsen et al., 2007; CSTE, 2017a). These include several measures of work-related injury risk, such as the number and rate of work-related hospitalization for severe traumatic injuries (Appendix D). Some fundamental state programs include targeted efforts focused on specific types of injuries. As described, a significant advantage of these state-based programs is the ability to identify and address local concerns working with other government agencies, policy makers, trade associations, unions, and community partners.

Workers' compensation data have been used extensively for surveillance of work-related injuries, including musculoskeletal disorders, in several states, particularly where a state agency is the sole insurer for workers' compensation, most notably, Washington State and Ohio. In 2015, NIOSH established a Center for Workers' Compensation Studies to promote the use of these data to improve workplace safety and health in additional states (see Chapter 6).

### Additional Surveillance Systems Relevant to Nonfatal Occupational Injuries

Some of the other surveys and data systems used by NIOSH to conduct surveillance of chronic disease (described in Table 4-3) also provide information on nonfatal and fatal occupational injuries. An advantage of such information that is collected within broader public health data systems is that, unlike employer-based reporting, it allows for the assessment of the contribution of work-related injuries to the overall injury burden in the United States. For example, a question about activity at time of injury in

**TABLE 4-3** Examples of Surveillance Systems for Nonfatal Occupational Injuries in the United States

| Surveillance System | Scope[a] | Type[b] | Responsible Agency(ies) | Condition(s) under Surveillance[c] |
|---|---|---|---|---|
| Survey of Occupational Injuries and Illnesses (SOII) | National,[e] 41 states and the District of Columbia | P | BLS* States* | O: Nonfatal work-related injuries (95%) and acute illnesses (5%) |
| National Electronic Injury Surveillance System—Occupational Supplement | National | P | NIOSH* CPSC* CDC | O: Work-related injuries treated in emergency departments |
| Targeted State-Based Injury Surveillance Systems | 3 states | P, C, varies by system | States* NIOSH* | O[g] |
| Severe Injury Reporting | National, State | C | OSHA* | O: Severe injuries (hospitalizations, amputations, eye loss) |

NOTES: Asterisk indicates funding agency. BLS, Bureau of Labor Statistics; CDC, Centers for Disease Control and Prevention; CPSC, Consumer Product Safety Commission; NIOSH, National Institute for Occupational Safety and Health; OSHA, Occupational Safety and Health Administration.

[a]Geographic levels at which findings are publicly available.

[b]P, population based: data are collected on a census or representative sample of a defined population and allows for assessing extent of a health-related event and monitoring trends with this population over time/locale; C, case based: focus of data collection is on individual cases that require follow-up or immediate public health action. These approaches are not mutually exclusive.

the core module of the National Health Interview Survey (NHIS) allowed researchers to estimate that 29 percent of all injuries among the working-age population occurred at work. Among employed men age 55 to 64 years of age, this figure was 49 percent (Smith et al., 2005). The National Center for Health Statistics (NCHS) is currently revising the NHIS and it is not clear whether this and other questions relevant to OSH surveillance will be retained in the core module (see further discussion of NHIS below).

Administrative databases originating in the health care system are yet another important source of information on work-related injuries but their

| Data Source(s) | Population Covered[d] | Approach | Time Frame for Report Release |
|---|---|---|---|
| Employer reports | Most workers[f] | Sample | Annual |
| Emergency department records | All workers<br>Special follow-back studies | Sample of 67 emergency departments nationwide | Intermittent |
| Multiple sources, vary by system | Varies by system | Census and some case series, varies by system | Varies by system |
| Employer reports | All workers covered by federal or state OSHA | Case series | Ad hoc |

[c]O, outcome; H, hazard; E, exposure.

[d]The population covered may include active and former workers, retirees, and others depending on the system.

[e]Information for Guam, Puerto Rico, and the U.S. Virgin Islands is collected and public but not included in national estimates.

[f]Excludes self-employed, workers on small farms, domestic workers, U.S. postal workers, and federal workers, approximately 9 percent of the U.S. workforce.

[g]Massachusetts, young worker injuries, and hospital worker injuries; Michigan, amputation, burns, crushing injuries, farm injuries, and skull fractures; Washington, trucking industry worker injuries.

use is currently limited by lack of information on work-relatedness in these records. Payment information in these data systems indicating workers' compensation as payer can be used to identify some but not all work-related cases, as not all workers are covered under workers' compensation, and some injured workers who are may not file claims or their claims may be denied (Spieler and Burton, 2012; Groenewold and Baron, 2013; Sears and Bowman, 2016). The *International Classification of Diseases, 10th Revision, Clinical Modification* (ICD-10-CM) includes a supplemental External Cause of Injury code—"External Cause Status"—that indicates if injury occurred while engaged in civilian activity for pay or income. ICD-10-CM was implemented in clinical settings in the fall of 2016 and the extent that these codes are being used in various health data sources remains to be evaluated.

**Conclusion: The use of the supplemental External Cause of Injury codes has substantial potential to enhance the utility of hospital inpatient, emergency department, and outpatient data systems for OSH surveillance.**

### Work-Related Musculoskeletal Disorders

NIOSH's prioritization of research and prevention resources based on burden, need, and impact directs attention toward the detrimental impact that work-related musculoskeletal disorders (WMSDs) have on workers and social insurance systems. Current estimates of the significant burden of WMSDs come from the BLS Survey of Occupational Injury and Illness—31% of all occupational injuries (BLS, 2016g), state-based surveillance programs usually using workers' compensation data—43 percent of compensable claims in Washington State (Marcum and Adams, 2017), and the National Health Interview Survey Occupational Health Supplement (CDC, 2017e). WMSDs likely strain the U.S. Social Security Disability Insurance (SSDI) System as the expiration of limited benefits provided for long-term disability under workers' compensation are "taken-up" by benefits offered under SSDI (Reville and Schoeni, 2004; O'Leary et al., 2012).

Throughout their history NIOSH and OSHA have recognized work as a risk to musculoskeletal health (NIOSH, 1997). A comprehensive review by the National Research Council and Institute of Medicine addressed the work-relatedness of musculoskeletal disorders (NRC, 2001). This review gave recommendations to BLS to revise their data collection and reporting systems for more comprehensive surveillance of WMSDs and recommended for NIOSH a lead role in "developing uniform definitions of musculoskeletal disorders for use in clinical diagnosis, epidemiologic research, and data collection for surveillance systems." Some of these recommendations to BLS and NIOSH are reiterated in this report, such as "including details on non-lost-workday injuries or illnesses (as currently provided on lost workday injuries) to permit tracking of these events in terms of the variables now collected only for lost workday injuries (age, gender, race, occupation, event, source, nature, body part, time on the job)" (NRC, 2001). The report also supported a broader surveillance approach recommending collection of additional information on the worker's use of tools and technology, the organizational conditions under which work is performed and development of exposure assessment tools which could be used for hazard surveillance.

Under current OSHA recordkeeping requirements, establishment level WMSD data are not easily available and OSHA's efforts to improve establishment level recordkeeping requirements identifying WMSDs on OSHA logs have not been successful (GAO, 2016). The OSHA log prior to 2001 had included a column for "repeated trauma cases," which was inclusive of some MSDs but not all (GAO, 2016). OSHA's 2001 revised recordkeeping regulation included a column to identify MSD cases, but was deleted in a 2003 amendment to the standard. In 2010, OSHA proposed to modify the Log 300 form to include a specific column to identify which injuries are MSDs, which would provide more complete information on the extent of MSDs both in the workplace and in the SOII, but due to opposition and intervention by Congress, this proposed addition has not been finalized.

In the second decade of the National Occupational Research Agenda (from 2006-2016; NORA), industry sector councils included WMSD surveillance and prevention as part of their research agendas. In both the second and the third upcoming decade of NORA, NIOSH has included a Musculoskeletal Disorders Cross Sector Council, which reinforces the ubiquitous, pervasive nature of risk factors for WMSDs in all sectors of the U.S. economy.

**Conclusion: Work-related musculoskeletal disorders represent a large physical and economic burden on the U.S. workforce demanding a continued emphasis on surveillance of these disorders.**

## OCCUPATIONAL DISEASE SURVEILLANCE

Surveillance of diseases in which an occupational hazard or hazards were the cause or contributing factor is especially challenging because the illnesses of interest (e.g., coronary artery disease, chronic obstructive pulmonary disease [COPD], lung cancer, and renal failure) can often have many potential risk factors, including work, that can contribute to disease development, and there is often substantial lag time between initial exposure and disease onset. Limited emphasis on occupational health in medical school curricula leaves most physicians ill equipped to assess work-relatedness of multifactorial conditions, which adds to the challenge (Michas and Iacono, 2008). The importance of work among the causes for these multifactorial conditions is underappreciated. Estimates of the preventable causes of these conditions generally ignore the impact of work resulting in incidence, prevalence, and mortality estimates that fail to identify primary prevention opportunities in the workplace. While the fraction of multifactorial illnesses that can be attributed to work may be small for common conditions such as lung cancer, workplace exposures may still account for a substantial human and economic burden. For example, consensus statements from the American Thoracic Society (ATS) based on the review of the medical literature have concluded that 15 percent of new-onset asthma among adults is caused by workplace exposures and that another 25 percent of adult asthma is work aggravated (Balmes et al., 2003; Torén and Blanc, 2009; Henneberger et al., 2011; Blanc, 2012). Similarly, the ATS and others have estimated that work exposures contribute to at least 15 percent of COPD cases (Hnizdo et al., 2002; Balmes et al., 2003; Raherison and Girodet, 2009). Other estimates of the attributable risk of work have been developed: 6.3 to 18 percent of coronary artery disease deaths, and 8.2 to 14.5 percent of chronic renal failure deaths (Steenland et al., 2003). A recent systematic review that covered 46 years and 8,000 published studies found good evidence for the role of a variety of chemical exposures in heart disease, pulmonary heart disease, stroke, and high blood pressure (SBU, 2017).

In considering surveillance of long-latency conditions, it is useful to distinguish between diseases (such as silicosis and coal workers' pneumoconiosis) that are almost invariably work related, and the diagnosis itself as an indicator for work-relatedness (pathognomonic occupational diseases) and multifactorial diseases for which occupational causes are only one of a number of contributing factors. Approaches to surveillance of the long-latency work-related pathognomonic diseases and common multifactorial diseases differ markedly. For pathognomonic diseases, data sources such as hospital discharge records that include diagnostic information can be used for surveillance to assess extent of the condition. Surveillance of multifactorial diseases is much more challenging. For conditions with known occupational etiology and a relatively high fraction attributed to work such as asthma or COPD, case-based approaches that involve follow-up with individuals or their providers to ascertain work-relatedness and sources of exposure are feasible. Also, surveillance of exposures can be a useful adjunct (see discussion of hazard and exposure surveillance below).

Surveillance of common multifactorial conditions, where the occupational contributions are less well studied but where the conditions are prevalent in the adult population (e.g., coronary artery disease, mental disorders) has generally been restricted to monitoring patterns of disease in relation to basic employment information (e.g., usual industry and occupation). Results from these efforts have been used to generate hypotheses about potential occupational associations that inform research priorities, and the approach may identify statistical aberrations (hot spots) that warrant follow-up. Research on cardiovascular disease (Fang et al., 2010; Costello et al., 2016; Du et al., 2016) and mental disorders (LaMontagne et al., 2008; Cohidon et al., 2012; Niedhammer et al., 2014; Theorell et al., 2015) has identified work-related risk factors that could be incorporated into surveillance efforts. For example, increased risk of cardiovascular disease among those exposed to fine particulates and of mental health problems among those working in stressful environments. In such efforts, disease and exposure information would need to be linked (see Chapter 6).

There are opportunities to advance knowledge about the importance of occupation as a contributor to many health conditions that have multifactor etiology with more extensive information about the employment history of individuals. The occupational medicine community has long advocated the collection of a more extensive occupational history in the individual patient's medical record. These efforts have focused primarily on physician education but with little success. The potential of obtaining occupational information systematically through the electronic health record (EHR; see Chapter 6) offers a new opportunity to expand understanding of the relationship of work to chronic health conditions. Once in the EHR, current occupational information and, more importantly, the development of full occupational histories will permit examination of specific common conditions to seek important signals of occupational factors. Once noted, these can be tracked, hot spots can be identified, and formal etiologic studies can be planned.

It has been long recognized that the SOII is not an adequate source of information on occupational illnesses. The SOII captures only a limited percentage of acute illnesses such as dermatitis and few of the long-latency occupational illnesses with some relationship to work (Windau et al., 1991; Luckhaupt et al., 2013a; Alarcon, 2016). Although the 1987 NRC report recommended that NIOSH take the lead in occupational disease surveillance and develop a comprehensive national surveillance system using a wide variety of data, this funding to do so has not been provided to NIOSH. Rather, over the past several decades, NIOSH has taken more limited steps to advance surveillance of occupational disease.

The major strategies used by NIOSH have been to leverage use of existing health data sources and surveys and to support state health agencies in building case- and population-based surveillance systems for select occupational health conditions. Key systems and approaches currently in place are listed in Table 4-4 and described briefly below.

### Death Certificate Data

Since 1980, NIOSH has partnered with up to 30 states and the NCHS to obtain death certificate data and to code the narratives in these records on usual industry and occupation according to the Census Bureau's Industry and Occupation classification system. The death certificate data from these states (including underlying and contributory causes of death, demographic data, and usual industry and occupation information) serve as the basis for the NOMS system which is used by NIOSH to monitor trends in chronic disease and other causes of mortality by occupation and industry. Occupational diseases such as asbestosis or silicosis that are almost always work related can also be characterized. NIOSH and other researchers have

**TABLE 4-4** Examples of Surveillance Systems for Occupational Illnesses in the United States

| Surveillance System | Scope[a] | Type[b] | Responsible Agency(ies) | Condition(s) under Surveillance[c] |
|---|---|---|---|---|
| National Health Interview Survey Periodic (5-year) Occupational Health Supplement | National | P | NCHS* NIOSH* | O, H<br>Multiple outcomes and hazards that vary with NIOSH supplement |
| Adult Blood Lead Evaluation and Surveillance Program | 26 states (and aggregated) | P, C | NIOSH* States* | E: Elevated blood levels |
| SENSOR Pesticide Surveillance | 12 states (and aggregated) | C, P | NIOSH* EPA* States | O: Occupational pesticide-related injury and illness |
| Work-related asthma surveillance | 5 states (and aggregated) | C, P | NIOSH* States | O: Asthma caused or exacerbated by work |
| Multisource silicosis surveillance | Michigan | C,P | NIOSH* State | O: Silicosis morbidity and mortality |
| Sharps injuries (blood-borne pathogen exposure) among hospital workers | Massachusetts | P | NIOSH* State* | E: Injuries due to contaminated needles and other sharp devices |
| National Occupational Mortality Surveillance (NOMS) | Aggregated data from 17 states[f] | P | NIOSH* NCHS* | O: Chronic disease deaths by industry and occupation |

| Data Source(s) | Population Covered[d] | Approach | Time Frame for Report Release |
|---|---|---|---|
| In-person household survey | Adult civilian population living in U.S.[e] | Sample | Intermittent |
| Clinical laboratory reports | All adults age ≥16 | Census | Annual |
| Multiple data sources | All workers | Case series | Intermittent |
| Multiple data sources: provider case reports, hospital, workers' compensation records, etc. | All workers | Case series | Annual |
| Multiple data sources: provider case reports, hospital workers' compensation records, death certificates | All workers | Census | Varies by state |
| Mandated sharps injury logs maintained by acute and chronic care hospitals | All workers in hospitals including students in training | Census | Annual |
| Death certificates | All workers | Census | Intermittent |

*continued*

**TABLE 4-4** Continued

| Surveillance System | Scope[a] | Type[b] | Responsible Agency(ies) | Condition(s) under Surveillance[c] |
|---|---|---|---|---|
| National Occupational Respiratory Mortality Surveillance (NORMS) | National and state, some county level data | P | NIOSH* NCHS* | O: Pneumoconiosis, hypersensitivity pneumonitis and mesothelioma deaths; other respiratory disease deaths by industry and occupation |
| Occupational Cancer Surveillance | 5 states aggregated[g] | P | NIOSH* NIOSH* | O: Cancer incidence by industry and occupation |
| Hearing loss surveillance | National | P | NIOSH* | O: Hearing loss |

NOTES: Asterisk indicates funding agency. NCHS, National Center for Health Statistics; NIOSH, National Institute for Occupational Safety and Health; OSHA, Occupational Safety and Health Administration.

[a]Geographic levels at which findings are publicly available.

[b]P, population based: data are collected on a census or representative sample of a defined population and allows for assessing extent of a health related event and monitoring trends with this population over time/locale; C, case based: focus of data collection is on individual cases that require follow-up or immediate public health action. These approaches are not mutually exclusive.

published studies based on analyses of these data that have both confirmed previous risks and pointed to new associations that require further research (Dubrow et al., 1987; Blair et al., 1993; CDC, 1995; Burnett et al., 1997; Savitz et al., 1998; Colt et al., 2001; Luckhaupt and Calvert, 2008; Robinson et al., 2015). Findings are periodically updated and made available on an interactive NIOSH website (Box 4-2).

Electronic death registration systems, now implemented in most states, provide NIOSH with the opportunity to include more states and provide more timely coded industry and occupation data (PHII, 2016). NIOSH is planning to conduct a pilot effort with NCHS in 2018 to apply computer-assisted coding tools to code all industry and occupation data collected in real time from the 17 states now participating in NOMS. If successful, this

| Data Source(s) | Population Covered[d] | Approach | Time Frame for Report Release |
|---|---|---|---|
| Death certificates | All workers | Census | Intermittent |
| State cancer registries | All workers | Census | Intermittent |
| Sample of audiometric service providers and U.S. Air Force | All workers | Case series | Peer-reviewed papers |

[c]O, outcome; H, hazard; E, exposure.
[d]The population covered may include active and former workers, retirees, and others depending on the system.
[e]Noninstitutionalized population.
[f]Number of participating states has varied over time.
[g]Previously one state; pilot under way expanding to five states.

partnership will provide the opportunity to analyze mortality patterns by industry and occupation in all 50 states.

The National Occupational Respiratory Mortality System (NORMS) is an interactive data system that is based on mortality data from death certificates provided to NIOSH annually by the NCHS (see Box 4-2). This national database contains information about deaths from 1968 forward with underlying or contributory causes of death from select respiratory conditions known to be associated with work: pneumoconiosis, hypersensitivity pneumonitis (since 1979), and mesothelioma (since 1999) (Figure 4-4). Counts, age-adjusted rates, and potential years of life lost for all U.S. residents and some states and counties by age, gender, and race are generated from the database for various time periods and included in NIOSH's electronic Work-Related Lung Disease Surveillance System (eWoRLD), an interactive web-based application that presents data on both occupational morbidity and mortality due to respiratory disease (NIOSH, 2017d; see Box 4-2). The national database does not include information about the usual industry and occupation of the decedent. To address this, data from states participating in NOMS (see above) are incorporated in the NORMS system and used to monitor mortality patterns of a broader range of respiratory diseases by industry and occupation. This "industry and occupation database" includes information on deaths due to chronic obstructive lung disease, asthma, influenza, lung cancer, mycobacterial infection, pleural plaques,

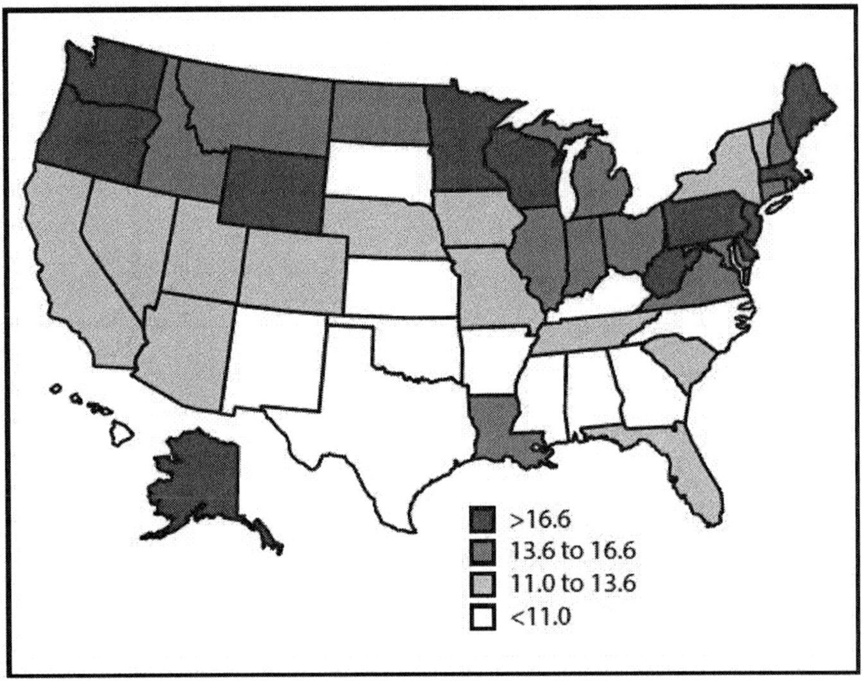


**FIGURE 4-4** Malignant mesothelioma annualized age-adjusted death rate* per 1 million population, † by state, United States, 1999-2015.
*Age-adjusted death rates were calculated by applying age-specific death rates to the 2000 U.S. standard population age distribution (CDC, 2017f). In two states (Maine and Washington), the age-adjusted death rate exceeded 20 per million per year.
†Decedents aged ≥25 years for whom the International Classification of Diseases, 10th Revision, codes C45.0 (mesothelioma of pleura), C45.1 (mesothelioma of peritoneum), C45.2 (mesothelioma of pericardium), C45.7 (mesothelioma of other sites), or C45.9 (mesothelioma, unspecified) were listed on death certificates were identified using CDC multiple cause-of-death data for 1999-2015.
SOURCE: Mazurek et al., 2017.

and tuberculosis in addition to the work-related respiratory diseases named above. Surveillance findings from this database, including proportionate mortality ratios for respiratory diseases by industry and occupation for several time periods, are likewise included in the eWoRLD. NORMS also includes tools for data users including crosswalks for comparing changes in the Census Bureau's Industry and Occupation Classification System codes over time (1990 and 2000 revisions) and Census population estimates used by the system in calculating mortality rates.

### Health Survey Data

The NHIS has been used by NIOSH and others to assess workers' health. The NHIS is an ongoing annual household survey of approximately 35,000 households and 87,500 individuals conducted by the NCHS (NCHS, 2016). The survey is designed to represent the civilian noninstitutionalized population residing in the United States at the time of the interview. Its primary objective is to obtain national estimates of health care utilization, health conditions, health status, insurance coverage, and access, as well as to monitor trends in illness and disability in the nation. The NHIS consists of a core set of questions that have remained relatively unchanged along with supplemental questions that vary. Historically the core questions have collected basic employment information (employment status, current occupation, and industry) so that NHIS data from every year could be used to study general trends in chronic disease and other health conditions by industry and occupation. NIOSH has periodically analyzed data from this general health section to provide baseline information on the health status of the workforce in major industry sectors (see Table 4-5). As mentioned above, the NHIS is currently undergoing revision and the most recent proposal is to collect industry and occupation information on a rotating basis (NCHS, 2017). This would substantially reduce the utility of the data for examining not only health outcomes and disability but health care access and utilization as well as work-related health behaviors.

Periodically NIOSH funds occupational health supplements (NHIS-OSH) to collect national-level data on various occupational health outcomes and exposures, focusing on information not available in the SOII (NIOSH, 2016d).

An advantage of collecting supplemental OSH data through the NHIS is that the data can be used to provide current national estimates on the prevalence of common work-related health conditions and exposures by industry and occupation (CSTE, 2013a). NIOSH also uses such data to assess progress in meeting the Healthy People 2020 objective concerning patient-physician communication regarding asthma and work (HHS, 2017). Additionally, NIOSH researchers have used the data for topical research,

**TABLE 4-5** Prevalence of Selected Health Conditions, Work Organization and Psychosocial Factors, and Physical and Chemical Exposures Among U.S. Workers (health care sector versus all U.S. workers, 2010), Occupational Health Supplement to the National Health Interview Survey

| | | All U.S. Workers | Health Care Sector |
|---|---|---|---|
| Prevalence (%) | | Prevalence (%) | Prevalence (%) |
| Health Conditions | In the past 12 Months[a] | | |
| | Dermatitis | 9.8 | 11.7 |
| | Carpal tunnel syndrome | 3.1 | 3.3 |
| | Current asthma | 7.2 | 8.7 |
| | Injury or poisoning at work | 2.8 | 3.6* |
| Work Organization Factors | Work ≥60 hours a week[b] | 7.2 | 5.3 |
| | Work ≥48 hours a week[b] | 18.7 | 14.4 |
| | Nonstandard work arrangements[a] | 18.7 | 10.7 |
| | Temporary employment[a] | 7.2 | 3.9 |
| | Nonstandard shifts[a] | 28.7 | 29.0 |
| Psychosocial Exposures | Job insecurity[c] | 31.7 | 22.8 |
| | Work-family imbalance[c] | 16.3 | 16.9 |
| | Hostile work environments[a] | 7.8 | 9.1 |
| Physical/Chemical Exposures | Exposure to potential skin hazards at work[a] | 20.6 | 25.7 |
| | Exposure to secondhand smoke at work[d] | 10.0 | 8.3 |
| | Exposure to outdoor work[a] | 24.7 | 6.9 |
| | Exposure to vapors, gas, dust, or fumes[e] | 25.0 | 14.2 |

[a]Among U.S. adults who have worked in the past 12 months.
[b]Among U.S. adults who have worked in the past 12 months, who only held one job.
[c]Among U.S. adults who were employed in the week prior to interview.
[d]Among nonsmoking U.S. adults who have worked in the past 12 months.
[e]Exposure during longest-held job (all other exposures refer to current or most recent job).
*These estimates have a relative standard error >30% and <50% and should be used with caution as they do not meet NCHS reliability/precision standards.
SOURCE: NIOSH, 2013.

including, for example, the congruence between current and usual occupation, and links between workplace mistreatment and sickness absenteeism (Luckhaupt et al., 2013b; Asfaw et al., 2014). The implementation of the OSH supplement on an ad hoc basis has been identified as problematic (CSTE, 2013a). There is a need for an ongoing NIOSH commitment to periodically incorporate the OSH supplement in the NHIS if this approach is to provide information about trends over time. With this recognition, NIOSH plans to support the supplement every 5 years; however, there is concern about funding as the survey costs are substantially increasing (Schnorr, 2016). Also, the wide range of potential questions related to occupational safety and health in light of the need to limit survey length raises challenges in consistently collecting data over time.

**Conclusion: Pending the outcome of the proposed BLS pilot household survey, the periodic administration of the OHS supplement would need to be re-evaluated.**

NIOSH is engaged in a number of additional activities to incorporate occupational information in existing health surveys and data systems. These are described in Chapter 6.

### Medical Monitoring Program Data

Several ongoing programs that routinely monitor worker health in specific industries provide data useful for occupational health and safety surveillance. The NIOSH Coal Workers' Health Surveillance Program is a medical monitoring program established by the Federal Coal Mine Health and Safety Act of 1969 (P.L. 91-173; NIOSH, 2017e). The intent is to prevent early coal workers' pneumoconiosis from progressing to a disabling disease. From 1971-2014, the program provided surveillance services to active underground coal miners that included occupational histories and chest radiographs classified for chances of pneumoconiosis by physicians with special training and certification by NIOSH (B readers[10]). New miners were required to participate in the program when first employed (preplacement) and at 3 years after hire, they were able to participate on a voluntary

[10]A "B reader" is a licensed physician who has passed a test classifying 125 chest films for the presence and severity of findings associated with pneumoconiosis, based on a standardized classification system maintained by the International Labour Organization (ILO). The B reader maintains certification by passing a test classifying 75 chest films for pneumoconiosis every 4 years. Although the testing/certification is part of regulations administered by NIOSH for providing medical examinations to coal miners, OSHA medical monitoring regulations for silica exposed workers require that a B reader classify the chest radiograph. For asbestos exposed workers, OSHA requires that a B reader, a board eligible/certified radiologist, or an experienced physician with known expertise in pneumoconiosis use the ILO system to classify the chest radiograph.

basis at about 5-year intervals thereafter. In 2014, the program expanded to provide services to active surface coal miners and to add spirometry and respiratory health questionnaires to surveillance services. NIOSH compiles the data to describe the prevalence of coal workers' pneumoconiosis and disseminates findings through a data query system (see Box 4-2). Despite the success in using these B-reader reports to conduct surveillance for coal workers' pneumoconiosis, no similar program has been set up for other common pneumoconiosis such as asbestosis or silicosis.

NIOSH has also encouraged large audiometry-testing firms to participate in a surveillance system for noise-induced hearing loss. These firms voluntarily submit audiometric hearing tests. The results have been published in peer-reviewed publications (Masterson et al., 2014). Results are analyzed by geographic regions and industry sector. NIOSH has used data from the National Health and Nutrition Examination Survey to examine the prevalence of hearing loss by occupation and industry (Tak et al., 2009); individual states have used the Behavioral Risk Factor Surveillance System (BRFSS) data (Stanbury et al., 2008) and case-based reporting by audiologists (MSU and MI DELEG, 2009).

OSHA requires medical monitoring of workers exposed to a number of hazards, including asbestos, noise, and silica. The asbestos and silica regulations require both chest radiographs and spirometry. The noise regulation requires audiometric testing. There is, however, no provision in these regulations to collect the results of the medical testing and use the data for surveillance of these conditions. In addition to mandatory medical monitoring programs, NIOSH, medical specialties and industry trade groups have developed recommended medical monitoring for individuals working with multiple substances, for example diacetyl and isocyanates.

**Conclusion: NIOSH has effectively used the results from a mandatory medical monitoring program for coal miners and a voluntary program for noise exposed workers for occupational illness surveillance.** Use of the results of mandatory medical monitoring required by OSHA and the recommended medical monitoring conducted by employers has the potential to improve surveillance data for selected occupational illnesses.

### Infectious Disease Surveillance

Infectious diseases can be an important public health problem in many different work settings (Molinari et al., 2007; Keech and Beardsworth, 2008; Edwards et al., 2016). Some work environments present increased risk of select infectious diseases. For example, livestock workers have been shown to be at risk of methicillin-resistant *Staphylococcus aureus* infections, hepatitis, Q fever, leptospirosis, and antibiotic-resistant *Escherichia coli* (Klous et al., 2016). Child care workers are at risk of several infectious diseases: cytomegalovirus through contact with toys or diaper changes, hepatitis B and C, and human immunodeficiency virus (HIV) transmitted through blood, and a variety of enteric pathogens such as hepatitis A, cryptosporidium, giardia, shigella, campylobacter, enteroviruses, and rotavirus transmitted by fecal-oral contamination through diaper changes or via sink faucets and the hands of child care workers or children (Reves and Pickering, 1992; Churchill and Pickering, 1997; Cordell et al., 2004). Health care workers can be exposed to several infectious agents through sharp injuries (e.g., HIV, hepatitis B virus, and hepatitis C virus; NIOSH, 2017f) as well as through direct patient care (e.g., pertussis and meningococcus) and the contaminated environment (e.g., *Clostridium difficile*) (Weber and Rutala, 2016). And both health care workers and corrections officers are at increased risk for tuberculosis (IOM, 2001; NIOSH, 2016e).

The workplace can also be a critical locus for transmission of infectious disease, regardless of initial causation. For example, the annual epidemics of influenza A that are tracked around the world impact workplaces of many types. Influenza can lead to pneumonia and dehydration and can worsen long-term medical conditions, such as congestive heart failure, asthma, or diabetes. Timely information about work-related transmission of infectious diseases can be strategically important to facilitate rapid assessment and intervention to control the spread of disease to other workers and the public, particularly workers in critical and public-related services such as health care, medical waste treatment, emergency response, postal and package delivery, utilities, and transportation.

In the United States, state, local, and territorial public health agencies take the lead on receiving cases of specified infectious conditions. The Council of State and Territorial Epidemiologists (CSTE) is responsible for defining and recommending which diseases and conditions are nationally notifiable and reported to the Centers for Disease Control and Prevention (CDC). All but five of the nationally notifiable diseases are infectious diseases. States voluntarily report these conditions to disease-specific CDC programs through the National Notifiable Diseases Surveillance System (NNDSS).

Of the 90 conditions included in the NNDSS, approximately 36 include some information on work largely because of one or more specific work relationships. As noted, information about work is important not only in identifying the exposure source for reported cases but to address the workplace as a locus of potential disease transmission. However, the occupational data collected currently are not harmonized across conditions nor coded using a standard coding system. CDC is undergoing an agency-wide effort both to harmonize the variables collected in its surveillance systems and to improve the NNDSS technological infrastructure by basing it on interoperable, standardized data and exchange mechanisms. CSTE has recommended that CDC incorporate industry and occupation and other work information as appropriate in CDC surveillance systems as feasible (CSTE, 2014). This ongoing work provides an important opportunity for NIOSH and its state partners to influence collection and harmonization of industry and occupation information for conditions within the NNDSS. There appears to be increasing recognition that standardized collection and coding of industry and occupation information could substantially enhance surveillance of infectious diseases in relation to work, a significant gap in OSH surveillance. NIOSH and CSTE are currently engaged with colleagues at CDC to identify next steps to be taken to move this forward. In a related effort, they are also collaborating in the development of a Reportable Condition Knowledge Management System, a system designed by CSTE as a web portal for public health agencies to manage reporting criteria to facilitate automatic electronic case reporting of reportable and notifiable conditions documented in electronic health records (CSTE, 2017b).

On another front, given the multiple hazards, including exposure to infectious disease, faced by heath care workers, NIOSH has developed the Occupational Health Safety Network (OHSN). OHSN is a web-based application to collect standardized data from employers on common work-related injuries and illnesses among hospital workers, which includes a module to collect information on injuries from contaminated needles and other sharp devices (sharps injuries). A convenience sample of 114 hospital systems (2014 data) is currently reporting data through OHSN. As part of its state-funded activity, NIOSH also contributes to the support of the Sharps Injury Surveillance System in Massachusetts. State law in Massachusetts requires hospitals to report case-level data on all sharps injuries to the state health department annually. Sharps injury surveillance is essentially an exposure surveillance system, as the focus is not on the injury or the development of an infection but the sharp injury as an indicator of potential exposure to blood-borne pathogens.

OSH surveillance programs in state health agencies may also collaborate with their infectious disease programs to investigate and address clusters of infectious disease in the workplace. For example, occupational safety and health staff worked with infectious disease colleagues in California to identify, investigate, and prevent coccidioidomycosis among construction workers in the solar industry (Wilken et al., 2015).

**Conclusion: Further work is needed to increase collaboration between infectious and occupational public health programs at the state and federal levels to ensure that information on work is regularly collected and considered as part of infectious disease surveillance activity, including investigations.**

### State-Based Surveillance Systems for Occupational Disease Surveillance

Approximately 10 states have built on mandatory state disease reporting requirements to implement case-based surveillance of selected occupational disease using a model initially developed by NIOSH—the Sentinel Event Notification System for Occupational Risks (SENSOR) (Rutstein et al., 1983; Baker, 1989). This model is based on the concept of a sentinel health event in which a single event is considered a sentinel or warning sign that the prevention system has failed and follow-up with the worker and/or the worksite may be warranted to prevent additional cases (Baker, 1989). Initially based primarily on case reporting by health care providers, the model has evolved over time to include use of multiple sources such as death certificates, administrative data sources such as statewide hospital data sets and worker's compensation records, laboratory reports, and other sources such as calls to poison control centers and agricultural extension programs for case ascertainment, e.g., silicosis (Schleiff et al., 2016). NIOSH identifies national priorities for targeted funding and states have the option of proposing their own targets. Case definitions for surveillance, which have been developed by NIOSH and the states, are used to confirm cases. Case follow-up may include public health investigations of the worksite, referrals to OSHA or other agencies to control exposures and protect other workers at risk, and steps to ensure the affected individual has appropriate medical treatment. Follow-up also allows for collection of additional data to better understand the epidemiology of the disease. Summary data are used by states to target broader-based prevention efforts and have included, for example, dissemination of prevention recommendations; changes in state policies; and educational outreach to employers, workers, and health care providers. While data from case-based surveillance may or may not be complete or representative, summary findings can identify new hazards and provide important, otherwise, unavailable information to inform prevention efforts (Rosenman et al., 2003).

Currently state-based surveillance systems are in place for surveillance of work-related lung disease, with a focus on asthma and silicosis, and pesticide-related injury and illnesses. All use multiple data sources for case ascertainment and data to inform prevention efforts. Participating states also submit their data to NIOSH for aggregation and analysis to gain a broader understanding of the problems, fill gaps in national level surveillance, and foster prevention activities such as those described in the following examples. Findings of work-related asthma associated with cleaning products in multiple states have led to changes in national policies regarding certification of products as "green cleaners" only if they contain no known asthma-causing chemicals (Green Seal Institutional Cleaning Products, 2017). Data from multiple states on illness associated with the use of pesticide-releasing foggers led the Environmental Protection Agency to issue new requirements for product labeling to improve user understanding of risks and safe use (CDC, 2008).

Additionally, 28 states participate in the NIOSH Adult Blood Lead Epidemiology and Surveillance (ABLES) program (2015 data). These states require clinical laboratories to submit reports of blood lead tests in both adults (and children) to a state health agency. Blood lead levels (BLLs) at or above the reference level of 5 μg/dL are considered an indicator of exposure (CSTE, 2015b). Laboratory reports generally do not contain information about the industry or occupation of the affected worker. States conduct follow-up of individual cases based on blood lead levels to ensure adequate medical treatment and removal from exposure and to identify the source of exposure, including industry of the affected worker, and to control exposures. Many state health agencies work with OSHA to conduct follow-up in workplaces where cases were exposed to lead. Summary data are used to monitor trends and to identify high-risk industries and communities for outreach. States participating in ABLES submit their data to NIOSH and summary findings are published annually (Alarcon, 2016). The NIOSH ABLES program, which provides resources for adult blood lead surveillance in most of the participating states, reports that the program has contributed to a greater than 60 percent decline in the prevalence of BLLs ≥25 *μ*g/dL among adults in the United States from 1994 through 2013 (NIOSH, 2016f) (see Figure 4-5) A limited number of states have developed similar laboratory-based surveillance systems for other metals such as cadmium and mercury (e.g., Michigan and New York; MI DHHS, 2015; New York State Department of Health, 2016).

The occupational health indicators generated by NIOSH-funded OSH programs in about half the states include measures of several diseases almost always caused by work such as mesothelioma and asbestosis as well as elevated blood lead levels in adults and occupationally related calls to poison control centers (Appendix D).

### Summary

In echoing the conclusions of the 1987 NRC committee, there are no modifications of the BLS employer survey that would enable it to measure the occurrence of occupational illnesses. Studies have documented that the SOII only captures a limited percentage of acute illnesses and is even less useful in counting long-latency occupational illnesses. Individual states and NIOSH collect occupational illness surveillance data but the data are not compiled and findings regularly released. Individual research publications, surveillance reports, and web applications are used to disseminate findings

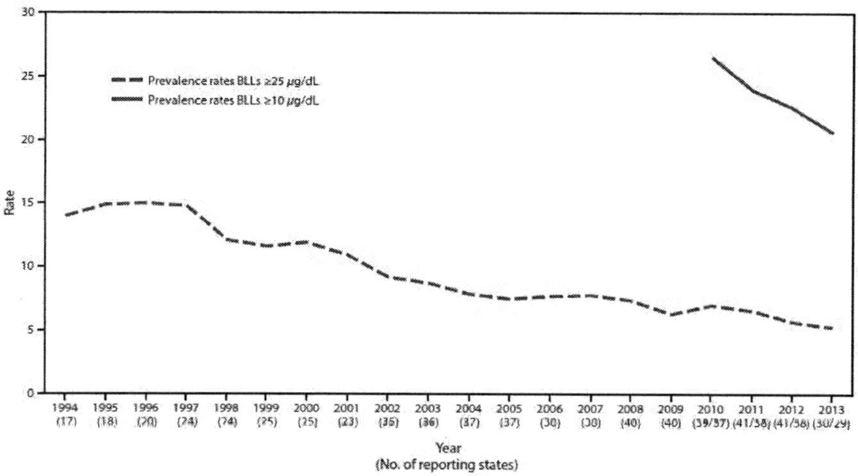


**FIGURE 4-5** National prevalence rate* of reported cases of elevated blood lead levels (BLLs),† by year (Alarcon, 2016).
*Per 100,000 employed adults aged ≥16 years. Denominator data extracted from 2015 U.S. Department of Labor, Bureau of Labor Statistics Local Area Unemployment Statistics (LAUS) program (http://www.bls.gov/lau/staadata.txt).
†Since 2009, the case definition for an elevated blood lead level is a BLL ≥10 µg/dL. For historical comparisons, prevalence rates at the previous case definition (BLL ≥25 µg/dL) are provided.
NOTE: A total of 30 states submitted data in 2013 (down from 41 states in 2012): Alabama, Alaska, Arizona, California, Colorado, Connecticut, Florida, Georgia, Illinois, Indiana, Iowa, Kentucky, Louisiana, Maryland, Massachusetts, Michigan, Minnesota, Missouri, Nebraska, New Jersey, New Mexico, New York, North Carolina, Oklahoma, Oregon, Pennsylvania, Vermont, Washington, Wisconsin, and Wyoming. Massachusetts provided data for BLLs ≥ 25 µg/dL. For 2013, the first number is the number of states reporting BLLs ≥ 25 µg/dL (i.e., 30 states in 2013), and the second number is the number of states reporting BLLs ≥ 10 µg/dL (i.e., 29 states in 2013).

but not with all occupational illnesses combined, not in conjunction with the release of the BLS SOII data, and not on a regularly scheduled basis.

**Recommendation B: NIOSH, working with the state occupational safety and health surveillance programs and across divisions within the agency, should develop a methodology and coordinated system for surveillance of both fatal and nonfatal occupational disease using multiple data sources.** The data should be analyzed, interpreted, and presented regularly in a comprehensive public report.

The data sources to be considered should include reporting by audiometric providers, disease registries (such as cancer and chronic renal failure), hospitals, laboratories, physicians, poison control centers, and health surveys as well as appropriate exposure databases. It is important that illness data collected by the states and NIOSH be analyzed and released in a timely manner. The data should be released in conjunction with BLS illness data in a manner that does not delay data released by BLS.

In the near term:

- NIOSH should combine information from the existing focused occupational disease surveillance systems (e.g., ABLES, pesticide illness, silicosis surveillance, and NORMS) and mesothelioma from cancer registries and other relevant occupational health indicators to provide a more comprehensive annual report on the extent of occupational illness morbidity and mortality that can be released in conjunction with information from the SOII. Methods for extrapolating from available data to generate national estimates should be explored.
- To enhance surveillance of occupational lung disease, NIOSH should require all B readers to report all chest radiographs interpreted to be positive for all types of pneumoconiosis.
- Increased collaboration between NIOSH and CDC infectious disease surveillance programs, with improved collection of occupational information, will be important to improve documentation of endemic and epidemic infectious disease related to work.

In the longer term:

- Gaps identified in the occupational illness surveillance system will need to be addressed through future developments that may involve
    - New or modified state regulations, requiring close coordination with the states, many of which have already promulgated reporting regulations.
    - Inference of occupational disease burden and trends that result from enhanced exposure assessment (Recommendation H, see Chapter 6).
    - Updating the list of occupational sentinel health events, establishing additional criteria for establishing a link between workplace exposures and common diseases.
- Action on recommendations that address the inclusion of occupational information in medical records (Recommendation J, see Chapter 7), federal health surveys and public health surveillance systems (Recommendation M, Chapter 7), and automated coding of the industry and occupa-

tion information (Recommendation L, see Chapter 7) will be important for ensuring the optimal implementation over time of this recommendation.

## OCCUPATIONAL HAZARD AND EXPOSURE SURVEILLANCE

Occupational hazard surveillance is the systematic assessment of the occurrence of workplace risks with potential to contribute to health, disease, and injury among working populations. Exposure surveillance measures actual risk by including distribution of, and the secular trends in exposure to workplace risks. In a public health context, hazard or exposure surveillance identifies settings or individuals exposed to hazardous levels of specific agents allowing for intervention to reduce risk. While there is no comprehensive occupational exposure surveillance system in the United States, the concept has been discussed over the past several decades and was included in the 1987 NRC report.

Risk is defined as the combination of hazard and exposure. A hazard is a substance or condition with an inherent ability to harm; for example, in the case of chemicals, the material's toxicity is the point of reference, while for radiant energy, the wavelength determines the hazard. Harm results when individuals or populations come into contact with, and are exposed to, the hazard. Both the intensity and the duration of exposure are important to determine. For some hazardous exposures the harm can just be acute (e.g., acute poisoning by carbon monoxide), for others the harms are cumulative or latent and cause or contribute to chronic disease (e.g., asbestos), and for others the harm may be both acute and chronic (e.g., lead).

The 1987 NRC report identified hazard and exposure surveillance as a priority and presented four recommendations:

- OSHA should include quantitative exposure information in its publicly available database.
- OSHA should require submission of all exposure measurements required by industry in complying with OSHA standards.
- NIOSH should compile and publish exposure data collected during health hazard evaluations as well as analyze these data to characterize the evaluated industries.
- NIOSH should include quantitative exposures in any future occupational hazard surveys.

Opportunities for creating systems to address these recommendations have been challenging; nevertheless, significant advances have been made. OSHA responded positively to the recommendations by including quantitative industrial hygiene measurements in their publicly available Integrated

Management Information System (IMIS), and these data are now being explored by researchers for quality and potential intervention research approaches. In addition, IMIS data are being used for surveillance purposes by NIOSH researchers that are focused on respiratory health and on surveillance. However, the specific goals and procedures for use of these data for ongoing surveillance have not been defined. In addition, NIOSH has compiled and created exposure-specific databases derived from the health hazard evaluation (HHE) investigations and to date has posted data for lead and noise (NIOSH, 2015b). Data on formaldehyde exposures are expected to be posted soon.

Large-scale hazard surveys, such as the National Occupational Hazard Survey (NOHS) and the National Occupational Exposure Survey (NOES) (see also Chapter 6), have not been conducted since the 1980s and those data, which were widely used, are now considered out of date and of limited utility (NIOSH, 1974; NIOSH, 2017g). In 2011, NIOSH conducted a web-based survey of health care workers—the Health and Safety Practices Survey of Healthcare Workers—to evaluate the extent of the use of well-known precautionary practices to minimize exposure to chemical hazards (NIOSH, 2017h). The primary lessons learned were that

- recommended exposure controls are not always implemented by the employer or employees;
- barriers to use of personal protective equipment (PPE) include that PPE is not provided or the perceived risk of exposure is underestimated, even for highly toxic chemicals such as chemotherapeutic drugs; and
- research is needed to evaluate factors preventing the use of safe handling practices.

Problems with a web-based survey were characterized and difficulty in reaching some of the health care worker component groups (housekeeping and environmental services) was noted. OSHA's publicly available IMIS and Chemical Exposure Health Data (CEHD) include quantitative industrial hygiene measurements from a variety of sources.[11] Assessment of IMIS data from the 1980s found that, despite the limitations in using exposure data from a compliance database, some surveillance objectives are met when examining exposures to airborne lead and airborne silica. An analysis of the airborne concentrations of lead collected during OSHA compliance in-

[11]The CEHD, available since 2010 as part of the OSHA Information System, also contains quantitative industrial hygiene measurements but these are limited to those that are processed by the OSHA Salt Lake Technical Center. The CEHD and IMIS have a significant degree of overlap (about 50 percent) but each data set contains a substantial amount of unique data (Lavoue et al., 2013a,b).

spections from 1979 to 1985 identified 52 industries which had more than one-third of the median air lead levels measured that were greater than the permissible exposure limit (Froines et al., 1990). The data developed in this analysis also indicated the need to investigate certain industries with high exposures but few inspections. Methods were also developed that permit ranking of potentially hazardous industries in a geographic area using IMIS data.

Researchers have continued to explore the effective use of IMIS data for risk characterization. A systematic review of studies using or analyzing IMIS data identified 29 such studies, most of which focused on single analytes such as lead or silica (Lavoue et al., 2013a). These authors also identified potential biases due to underreporting of values under the limit of detection. However, despite the limitations of these data, their potential for ongoing hazard characterization was highlighted. Additional modeling of the IMIS data using ancillary information about the workplace and its history of inspection demonstrated additional utility of the IMIS data in understanding some workplace characteristics associated with higher exposure levels (Sarazin et al., 2016). A number of other studies examining specific issues, for instance isocyanate asthma, have used IMIS data to identify and characterize high-risk conditions (Lefkowitz et al., 2015).

Despite the above examples using IMIS data, the limitations of using compliance data for surveillance are well recognized. The data are not comprehensive or balanced with respect to the exposures characterized, the workplaces or jobs selected for monitoring, or the complete reporting of results. In addition, although making these data available to the public is a big step forward, they are not routinely disseminated in a simple way for use by the public or health agencies. Promising developments in the surveillance of occupational hazards and exposures is further discussed in Chapter 6.

## CROSS-CUTTING ISSUES

This section provides a brief overview of several topics relevant across the different end points under surveillance: the role of state programs, surveillance research, and sources of information on populations at risk.

### State-Based OSH Surveillance Programs

The 1987 NRC panel recognized the importance of the state-federal partnership and recommended that NIOSH continue and expand its efforts to develop the occupational health surveillance capability of state health departments through technical assistance and financial support. NIOSH responded positively to these recommendations and, even in the absence of additional dedicated federal funding for surveillance, now supports some level of OSH surveillance activity in 27 of the states.[12] Initially, in the late 1980s, state activities focused largely on exploring use of death certificate and cancer registry data to examine patterns of disease in relation to employment characteristics. Today, states are exploring many different data sources, in some instances as part of multisource surveillance systems, to meet state surveillance objectives and to inform action to improve worker safety and health, while helping to fill gaps in national surveillance.

There are many recognized benefits: the state programs have access to unique state data sources; they are in a position to conduct case- as well as population-based surveillance with the attendant responsibility to follow up in individual workplaces; and, because the states have legal access to key identifiers, they are able to link data sources to develop a more comprehensive understanding of the magnitude and distribution of the conditions under surveillance. The states have also demonstrated the critical position they are in to use the surveillance findings to promote practical actions to improve worker safety and health.

While important advances have been made in OSH surveillance at the state level, there are significant obstacles to building effective programs across the nation. As described in Chapter 3, worker safety and health is seen largely as a responsibility of the Department of Labor. As a result, occupational health and safety and OSH surveillance have remained relatively low priorities in the general public health community at both national and state levels. Occupational safety and health has not received strong support from the U.S. Department of Health and Human Services (HHS) or CDC, and OSH surveillance has not been funded by CDC as a core public health function. Consequently, whereas in some public health domains, such as infectious disease and cancer prevention, surveillance programs are in most if not all states, close to half of the states report having little or no OSH surveillance capacity. In a nationwide assessment of state health agencies regarding epidemiologic and surveillance capacity in eight public health domains, only 20 percent of states reported having substantial epidemiologic and surveillance capacity in occupational health, the lowest percentage of all domains assessed (CSTE, 2013b). This results not only in lack of attention to occupational safety and health but missed opportunities for collaboration across public health domains to address convergent public health concerns that affect workers as well as the general public.

Given limited resources for OSH surveillance, states with OSH pro-

[12] Twenty-six states have funding from NIOSH for fundamental or expanded surveillance (detailed later in the chapter). Additionally, 5 states have NIOSH funding for work with workers' compensation data (of these 1 state is not part of the 26 states with fundamental or expanded surveillance).

grams rely on transient federal funding allocated through a highly competitive application process. While this is not unique to occupational health, it substantially hinders continuity and strategic consistency (Burkom, 2017). State programs come and go over funding cycles leaving gaps in the nation's OSH surveillance capacity. This problem is compounded by lack of a comprehensive national strategic plan for state-based surveillance identifying critical national priorities for expanded state-based surveillance. An additional "operational" barrier faced by states is the funding mechanism used by NIOSH to award funds for state OSH surveillance programs. In other public health domains, state-based surveillance activities are funded through applied public health practice mechanisms. However, NIOSH supports state surveillance activities through a competitive research application process (research cooperative agreements). This research application process introduces practical challenges for states. These include, for example, the inability of inexperienced states who wish to initiate OSH surveillance programs to meet research grant eligibility requirements such as innovation and the need for principal investigators with published research articles. For even experienced states, this has created hurdles—such as evaluation for ongoing funding based on innovation and the number of peer-reviewed publications—rather than success in collecting and using surveillance data to inform policy and practice. Additionally, these research applications require human-subjects review, which state institutional review boards are reluctant to conduct for ongoing surveillance programs, considered as applied public health practice.

**Conclusion: State OSH programs play an essential role in a national OSH surveillance system and are especially well positioned to disseminate surveillance findings to those in positions to take action to protect worker safety and health.** It is not necessary to conduct in-depth expanded programs for all conditions in all states in order to have an effective national surveillance system. However, contribution of such programs would be increased if implemented within the context of a comprehensive strategic surveillance plan developed by NIOSH in conjunction with the states as well as BLS and OSHA. Also, enhanced coordination among federally funded OSH and other public health programs at the state level have the potential to increase the effectiveness and efficiency of these state-based programs.

**Recommendation C: NIOSH should lead a collaborative effort with BLS, OSHA, the states, and other relevant federal agencies to establish and strengthen state-based OSH surveillance programs.** This should be carried out as part of a national coordinated effort to monitor priority conditions, hazards, and exposures; to identify emerging workplace risks; and to facilitate prevention programs that address these concerns. Furthermore, this should be carried out with the full support of and assistance from other parts of HHS and CDC.

In the near term:

- *OSH Agency Collaboration Within States:* NIOSH, BLS, and OSHA should actively encourage and promote collaboration among their programs in the states to reflect the national commitment to interagency effectiveness for OSH surveillance and leverage surveillance and prevention expertise across agencies. This should include sharing data and taking advantage of unique state-level data sets and case-based surveillance capacities to identify and respond to emerging occupational safety and health hazards and conditions.
- *Public Health Agency Collaboration Within States:* NIOSH and other CDC centers that support state-based surveillance and prevention activities should promote collaboration among their state-level programs to monitor and address public health problems of shared concern, such as violence, asthma, infectious disease, traffic safety, and health inequities among vulnerable population groups.
- NIOSH should also
    - o Explore and implement, as appropriate, alternative approaches to funding ongoing surveillance in the states as applied public health programs rather than research programs.
    - o Foster increased coordination and communication between its intramural and extramural surveillance programs.
    - o Encourage NIOSH-funded Education Research and Training Centers and Agricultural Health and Safety Centers to provide technical and research support to state surveillance programs in their regions as part of their required outreach and education core activity.

## Surveillance Research

Research is the process of systematic investigation to generate new facts, and it is important for all types of surveillance, including OSH surveillance. There are important distinctions, however, between research and surveillance (CDC, 2010). The most obvious distinction is that the primary goal of surveillance is to inform public health action, not necessarily to generate new knowledge. The dissemination of information produced by a surveillance system to guide action would be a central consideration in developing and evaluating surveillance systems. In contrast, research findings are usually disseminated through a process called knowledge translation, tending to be less well-structured and often ad hoc.

Another important distinction is that surveillance is an ongoing process, whereas research is not necessarily so. Accordingly, surveillance is usually funded through the core operating funds of a public health organization, while research is usually funded through competitions, which may be targeted (i.e., strategic) or investigator initiated. Despite these differences, it can be difficult to make the distinction between surveillance and surveillance research. This can happen, for example, when states conduct surveillance to protect the health of the state population, but also analyze the state data together with a national data set to generate new knowledge.

Surveillance research can generally be divided into two broad categories. The first is research that *uses data collected through surveillance* to generate new knowledge. Generating data for research is a well-accepted objective for a surveillance system, although not the primary objective for most systems. The second type is research to develop or adapt methods for surveillance. This type of applied research is less common, although increasingly recognized as being important for improving the efficiency and effectiveness of all types of public health surveillance. Applied surveillance research will play an important role in developing new approaches and adapting new technologies to build a smart OSH surveillance system for the 21st century.

### *Research Using Surveillance Data to Generate New Knowledge About Occupational Health*

Within a surveillance system, the data collected are usually limited to the minimum data necessary to accomplish the objectives. This limitation is imposed for multiple reasons, including cost, protection of privacy, and acceptability of the system by those who provide data. When supporting research is an objective of a surveillance system, then the scope of data collection may be greater than it otherwise would be (i.e., if supporting research were not an objective). In any event, even if supporting research is not an explicit objective of a surveillance system, the data collected by the system can usually support research. Examples range from descriptive analyses, hypothesis-driven research, and research to assess the impact of interventions, including more general policies both as natural experiments and as simulated scenario analysis ("what if").

Given that research occurs under a different legal and ethical framework than public health surveillance, an important consideration in this type of research is to ensure that the researcher has the legal authority to use the data, that the necessary ethical approvals have been obtained, and that the necessary mechanisms are in place to ensure the protection of privacy. In practice, the same people conducting surveillance may also be conducting research, but this does not diminish, and may even augment, the need to clearly distinguish between the use of surveillance data for public health practice and research.

*Research to Develop and Evaluate Methods for Occupational Health Surveillance*

Applied research aims to develop, implement, and evaluate methods and tools for conducting surveillance. It is considered applied because the focus is to improve aspects of the surveillance process. The following are examples of applied research in occupational safety and health:

- Development of novel technologies for measurement of exposure (e.g., accelerometers in helmets, clothing to measure repetitive movements), and collection of outcomes (e.g., social media, crowd sourcing);
- Development of data-coding technologies (e.g., natural language processing and statistical methods) to be shared;
- Application of or development of new statistical methods in occupational health for detection of events, clusters, or other outliers;
- Linking in data sets to join variables, to assess reporting, to get denominator for rates, and to triangulate;
- Assessment of data quality and new data sources for case ascertainment;
- Estimation of attributable fraction of COPD caused by exposures at work; and
- Evaluation of whether traditional approaches to capturing and coding occupation and industry cover new work arrangements and development of new approaches to capturing this information.

To be research, the results or methods need to be generalizable. Thus, periodic evaluation of surveillance data may be research if the findings are generalizable while periodic evaluation of a surveillance system would be part of ongoing surveillance programming.

*Current Surveillance Research*

BLS intramural researchers conduct surveillance research both using CFOI and SOII data and are developing new methods. These include, for example, development of coding software for assigning standard codes for occupation and event based on free text, and web-sweeping approaches for ascertaining fatal occupational injuries in publicly available data sets. As described above, the agency has processes in place for researchers to apply to use CFOI and SOII data for research purposes and is in the process of making these data available at their Federal Statistical Research Data

Centers (in a manner that preserves data confidentiality). In recent years, BLS has also supported a program of intramural and extramural research specifically aimed at better understanding the undercounting of injuries and acute illnesses in the SOII and factors that contribute to it. As will be described in Chapter 6, they are also supporting research to explore the feasibility of a worker survey.

NIOSH likewise conducts and supports surveillance research using surveillance data as well as applied research to develop new surveillance methods. Within NIOSH, investigators periodically compete with other researchers for funds to support investigator-initiated surveillance research projects using existing data sets. NIOSH also funds investigator-initiated research activities carried out by extramural researchers in universities and the states and other nonprofit organizations (for examples, see NIOSH, 2014b). With the exception of an extramural funding stream specifically designated for surveillance research in 2000, extramural applications for surveillance research are submitted to the general NIOSH extramural research funding competition. NIOSH does not appear to have a current research agenda for surveillance, although many of the NORA industry sectors have identified surveillance activities among their research goals.

### Sources of Information on Populations at Risk

This chapter thus far has focused primarily on approaches used to collect data on health outcomes or hazards (numerator data). As mentioned in the 2009 Institute of Medicine (IOM) report *Traumatic Injury Research at NIOSH*, information about the "population at risk (denominator data) is also critical for surveillance and necessary to calculate injury and illness rates that allow for identification of disproportionate risks among segments of the population and subsequent priority setting" (IOM, 2009).

The SOII and some state workers' compensation insurance programs collect data on numbers of workers and hours of work as part of ongoing data collection which allows for calculating injury and illness rates by industry (but not occupation and demographic characteristics such as age, gender, and race and ethnicity.) Other systems, such as ABLES, CFOI, and NEISS-Work, rely on external sources of denominator data to use in calculating incidence rates by employment and demographic characteristics. A number of different sources of employment data are collected by federal statistical agencies, each with strengths and limitations, and the choice depends upon the availability of robust data at the geographic level of interest and intended purpose. An important distinction is that between employee count and hours-based incidence rates. The denominator for hours-based incidence rates, which takes into account time persons are at risk, is generally expressed as a full-time equivalent (FTE).[13] As reported by BLS, count-based incidence rates underestimate risk among part-time workers, an important consideration in addressing the need to understand the injury and illness experience of the increasing numbers of workers in contingent employment situations (Ruser, 1998). BLS publishes hour-based rates in both the SOII and in CFOI (BLS, 2010).

One of the most widely used sources of employment information in generating injury or illness rates at the national and state levels is the Current Population Survey (CPS) conducted by BLS. BLS maintains a public use access file and NIOSH has developed a user-friendly query system—the Employed Labor Force query system—based on these data that provides employed worker population estimates of both counts of workers and FTEs (workers age 15 years or older) from 1980 through the present. A significant advantage of the CPS employment data is that this is a panel survey and data on employment and hours of work are collected from the same individual over time and aggregated to generate annual employment estimates. A disadvantage is that the sample size is often too small to generate robust rates for detailed segments of the population at the state or local level. The American Community Survey (ACS), conducted by the U.S. Census Bureau, is another important source of employment estimates with the advantage of a larger sample size. A disadvantage is that information about weeks and hours of employment is collected from individuals at a single point in time and thus subject to recall bias or error due to changes in work and work schedules over the course of the year. Also, the ACS reports employment data as a categorical rather than a continuous variable and offers no recommended method for computing FTEs. The Quarterly Census of Employment and Wages is yet another source of employment information. Unlike both the CPS and ACS, which are household surveys, the Quarterly Census of Employment and Wages Program at BLS provides data on the numbers of employed workers by industry provided by employers as part of the unemployment insurance system. This source may be expected to provide the most accurate count of workers but information about occupation hours of work or demographic characteristics is not provided. The BLS Current Employment Survey is another source of employment data based on data provided by a smaller sample of employers on a more frequent basis. BLS also generates monthly Local Area Unemployment (and employment) statistics for census areas and metropolitan regions combining data from employer-reported and household survey sources. Unemployment

[13]An FTE is the number of hours worked by one employee on a full-time basis. The concept is used to convert the hours worked by several employees into hours worked by full-time employees. A full-time employee in generally considered to work 40 hours per week and 50 weeks per year, or 2,000 hours per year.

Insurance (UI) wage files maintained by states for purposes of determining eligibility of claimants for UI benefits are another source of employment data for potential use in OSH surveillance.

NIOSH is currently working in collaboration with state and academic partners to explore the use of alternative data sources alone and in combination for generating employment denominators for use in OSH surveillance. This important research needs to be conducted. Practical guidance tools for selecting appropriate denominators for state OSH surveillance programs and other researchers are also needed. With changes in the health care delivery system there is increasing emphasis on documenting health needs at the community level. For example, under the Patient Protection and Affordable Care Act (P.L. 111-148) all nonprofit hospitals are required to conduct community health needs assessments to qualify for Medicare and Medicaid reimbursement (IRS, 2016). Local health departments also prepare a community health assessment as part of the public health department accreditation process (PHAB, 2015). Alternatives for generating community-level OSH surveillance information need to be explored.

Accurate documentation of employment among the contingent workforce in light of the changing nature of work and work relationships is a significant challenge and has potentially important implications for generating estimates of health and safety risk. Between 1995 and 2005, BLS collected data on contingent and alternative employment arrangements in periodic supplements to the CPS (GAO, 2015). These supplements were discontinued due to insufficient funds. While BLS has often requested funding to repeat this survey every 2 years, Congress has not allocated funding to do so. BLS did field a contingent worker supplement again in 2017 with one-time DOL funding, but it is unclear when the survey will be repeated. These surveys are needed to provide critical information about the workforce. Additional research is also needed to assess the validity of information on alternative work arrangements collected in population surveys that has implications for OSH surveillance. Research on how to generate authoritative labor statistics in light of the many changes in the structure of work including the emergence of the on-demand economy is a critical area of research with relevance far beyond OSH surveillance (NASEM, 2017).

Another concern expressed by OSHA to the committee is the lack of comprehensive data on establishments in the United States by industry that could be used to inform outreach and dissemination efforts. Although this information is collected through the unemployment insurance system overseen by BLS, it is not available to OSHA. Public health access to unemployment insurance listings of establishments by North American Industry Classification System code is also limited and varies by state.

**BOX 4-4**
**Key Gaps in Occupational Safety and Health Surveillance**

**Objective: Guide immediate action to address urgent health events**
*Gap:*
- Lack of a national approach for information sharing and collaboration among agencies to allow for timely identification and response to serious new and emerging health conditions and hazards

**Objective: Measure the burden**
*Gaps:*
- Lack of nonfatal injury data on some major working populations including, the self-employed, workers on small farms, domestic workers, and contingent worker populations (e.g. temporary, gig, immigrant workers)
- Limited data on nonfatal injuries in populations under surveillance due to underreporting, insufficient collection of detailed information for majority of cases reported, limited indictors of injury severity
- Insufficient data on chronic occupational disease and limited synthesis of available information
- No approach for systematic collection of information about hazards and exposures
- Lack of data on economic and social impact of occupational injuries and illnesses

**Objective: Detect new or emerging hazards**
*Gaps:*
- Limited capacity in federal agencies to analyze available injury, illness, fatality, and exposure data to identify workplaces or populations at high risk, emerging trends, and actively promote use of data for prevention

## SUMMARY

While substantially improved since 1987, the current state of OSH surveillance continues to have gaps and lacks maturity across the end points under surveillance. A robust system for fatal injuries is in place. Advances have been made in surveillance of nonfatal injuries with some critical omissions or shortcomings remaining. There continues to be limited surveillance of most chronic occupational diseases. There is very limited surveillance of hazards and exposures necessary for informing effective approaches to prevention for long latency health outcomes. There is not a clear delineation of specific objectives for each component of the current surveillance system, and a comprehensive synthesis of findings across these systems is

**Objective: Identify high-risk industries, occupations, worksites and populations**
*Gaps:*

- Ready access to establishment level data by government agencies for targeting preventive outreach and enforcement
- Lack of data on nonfatal injuries and acute illnesses at the firm level and on multiemployer worksites
- Limited data on socio-demographic characteristics and work arrangements needed to identify potential vulnerable populations
- Lack of occupational information in data sources originating in the health care delivery system that are a crucial complement to employer-based reporting
- Insufficient employment data on the contingent workforce needed to generate estimates of risk

**Objective: Guide planning, implementation, and evaluation of prevention programs at establishment and state levels**
*Gaps:*

- No systematic means to foster use of data by employers and workers at the establishment level or to provide feedback on injury and illness experience to assist with hazard identification and prevention efforts
- Limited or no OSH surveillance capacity in many states and missed opportunities for collaboration across public health domains

**Objective: Generate hypothesis and make data available for research**
*Gap*:

- No surveillance research agenda

lacking. Although there is some coordination across the multiple entities engaged in surveillance, the current "system" operates more as a collection of separate and sometimes fragmented data systems, rather than a coordinated national OSH surveillance system that effectively promotes use of the data for prevention.

Using the framework of the objectives of an ideal surveillance system outlined in Chapter 2 the committee identified the major gaps in surveillance (see Box 4-4). Agencies are well aware of current limitations and are engaged in a number of ongoing activities with potential to address many of these gaps. These are presented in Chapters 6 and 7.

## REFERENCES

AFL-CIO. 2017. *Death on the job: The toll of neglect*. Available online at https://aflcio.org/sites/default/files/2017-04/2017Death-on-the-Job.pdf (accessed June 15, 2017).

Alarcon, W. A. 2016. Elevated blood lead levels among employed adults—United States, 1994–2013. *Morbidity and Mortality Weekly Report* 63(55):59-65.

Asfaw, A. G., C. C. Change, and T. K. Ray. 2014. Workplace mistreatment and sickness absenteeism from work: Results from the 2010 National Health Interview Survey. *American Journal of Industrial Medicine* 57(2):202-213.

Azaroff, L., C. Levenstein, and D. H. Wegman. 2002. Occupational injury and illness surveillance: Conceptual filters explain underreporting. *American Journal of Public Health* 92(9):1421-1429.

Azaroff, L., H. M. Nguyen, T. Do, R. Gore, and M. Goldstein-Gelb. 2011. Results of a community-university partnership to reduce deadly hazards in hardwood floor finishing. *Journal of Community Health* 36(4):358-368.

Baker, E. L. 1989. Sentinel Event Notification System for Occupational Risks (SENSOR): The concept. *American Journal of Public Health* 79(Suppl):18-20.

Balmes, J., M. Becklake, P. Blanc, P. Hennenberger, K. Kreiss, C. Mapp, D. Milton, D. Schwartz, K. Toren, and G. Viegi. 2003. American Thoracic Society statement: Occupational contribution to the burden of airway disease. *American Journal of Respiratory Critical Care Medicine* 167:787-797.

Bhandari, R., S. M. Marsh, A. A. Reichard, and T. R. Tonozzi. 2016. Characterizing emergency department patients who reported work-related injuries and illnesses. *American Journal of Industrial Medicine* 59(8):611-621.

Blair, A., M. Dosemeci, and E. F. Heineman. 1993. Cancer and other causes of death among male and female farmers from twenty-three states. *American Journal of Industrial Medicine* 23:729-742.

Blanc, P. D. 2012. Occupation and COPD: A brief review. *Journal of Asthma* 49(1):2-4.

BLS (Bureau of Labor Statistics). 2010. *Census of fatal occupational injuries-hours-based rates*. Available online at https://www.bls.gov/iif/oshnotice10.htm (accessed November 8, 2017).

BLS. 2012. *Occupational Injury and illness classification manual Version 2.01*. Available online at https://www.bls.gov/iif/oiics_manual_2010.pdf (accessed November 8, 2017).

BLS. 2015. *Occupational injuries and illnesses: A pilot study of job-transfer or work-restricted cases, 2011-2015*. Report 1056. Available online at https://www.bls.gov/iif/oshwc/osh/case/djtr2012.pdf (accessed November 8, 2017).

BLS. 2016a. *Scope of the census of fatal occupational injuries*. Available online at https://www.bls.gov/iif/cfoiscope.htm (accessed June 12, 2017).

BLS. 2016b. *Fatal occupational injuries in 2015, charts*. Available online at https://www.bls.gov/iif/oshwc/cfoi/cfch0014.pdf (accessed May 1, 2017).

BLS. 2016c. *Survey of Occupational Injuries and Illnesses (SOII)—Information for respondents*. Available online at https://www.bls.gov/respondents/iif (accessed April 4, 2017).

BLS. 2016d. *2015 Survey of occupational injuries & illnesses summary estimates charts package*. October 27, 2016. Available online at https://www.bls.gov/iif/oshwc/osh/os/osch0057.pdf (accessed September 18, 2017)

BLS. 2016e. *Employer-reported workplace injury and illness summary*. Available online at https://www.bls.gov/news.release/osh.nr0.htm (accessed June 15, 2017).

BLS. 2016f. *Table 2. Numbers of nonfatal occupational injuries and illnesses by case type and ownership, selected industries, 2015*. Economic News Release. Available online at https://www.bls.gov/news.release/osh.t02.htm (accessed June 13, 2017).

BLS. 2016g. *Nonfatal occupational injuries and illnesses requiring days away from work, 2015*. Economic News Release. USDL-16-2130. Available online at https://www.bls.gov/news.release/osh2.nr0.htm (accessed November 8, 2017).

BLS. 2017a. *Census of Fatal Occupational Injuries (CFOI)—Current and revised data*. Available online at https://www.bls.gov/iif/oshcfoi1.htm (accessed April 4, 2017).

BLS. 2017b. *Non-fatal occupational injuries and illnesses with days of job transfer or restriction, Survey of Occupational Injuries and Illnesses*. Available online at https://www.bls.gov/iif/oshwc/osh/case/osch0059.pdf (accessed November 8, 2017).

BLS. 2018. *Injuries, illnesses, and fatalities databases*. Available online at https://www.bls.gov/iif/data.htm (accessed January 31, 2018).

Boden, L. I., and A. Ozonoff. 2008. Capture-recapture estimates of nonfatal workplace injuries and illnesses. *Annals of Epidemiology* 18(6):500-506.

Burkom, H. S. 2017. Evolution of public health surveillance: Status and recommendations. *American Journal of Public Health* 107(6):848-850.

Burnett, C., J. Maurer, H. M. Rosenberg, and M. Dosemeci. 1997. *Mortality by occupation, industry, and cause of death, 24 reporting states (1984–1988)*. Cincinnati, OH: NIOSH. DHHS (NIOSH) Document No. 97-114. Available online at https://www.cdc.gov/niosh/docs/97-114/pdfs/97-114.pdf (accessed May 5, 2017).

Byler, C. 2013. Hispanic/Latino fatal occupational injury rates. *Monthly Labor Review* February:14-23. Available online at https://www.bls.gov/opub/mlr/2013/02/art2full.pdf (accessed April 4, 2017).

CDC (Centers for Disease Control and Prevention). 1995. Proportionate mortality from pulmonary tuberculosis associated with occupations—28 states, 1979–1990. *Morbidity and Mortality Weekly Report* 44:14-19.

CDC. 2008. Illnesses and injuries related to total release foggers: Eight states, 2001-2006. *Morbidity and Mortality Weekly Report* 57(41):1125-1129.

CDC. 2010. *Distinguishing public health research and public health nonresearch*. Available online at https://www.cdc.gov/od/science/integrity/docs/cdc-policy-distinguishing-public-health-research-nonresearch.pdf (accessed June 15, 2017).

CDC. 2016. Strategies for reducing health disparities; selected CDC sponsored intervention, US, 2016. *Morbidity and Mortality Weekly Report Supplement* 65(1). Available online at www.cdc.gov/mmwr/volumes/65/su/pdfs/su6501.pdf (accessed November 8, 2017).

CDC. 2017a. *The work-related injury statistics query system*. Available online at https://wwwn.cdc.gov/wisards/workrisqs/ (accessed April 5, 2017).

CDC. 2017b. *Employed Labor Force (ELF) query system*. Available online at https://wwwn.cdc.gov/wisards/cps/ (accessed December 1, 2017).

CDC. 2017c. *Occupational injury and illness classification system*. Available online at https://wwwn.cdc.gov/wisards/oiics/ (accessed December 1, 2017).

CDC. 2017d. *Occupational injury and illness classification system: Coding selection rules*. Available online at https://wwwn.cdc.gov/wisards/oiics/ (accessed December 1, 2017).

CDC. 2017e. *NIOSH worker health charts: NHIS occupational health supplement. 2017*. Available online at https://wwwn.cdc.gov/Niosh-whc/source/ohs (accessed November 8, 2017).

CDC. 2017f. *Multiple cause of death 1999-2005*. Available online at https://wonder.cdc.gov/wonder/help/mcd.html#Age-Adjusted Rates (accessed December 1, 2017).

Churchill, R. B., and L. K. Pickering. 1997. Infection control challenges in child-care centers. *Infectious Disease Clinics of North America* 11(2):347-365.

Cohidon C., G. Santin, J. Chastang, E. Imbernon, and I. Niedhammer. 2012. Psychosocial exposures at work and mental health potential utility of a job-exposure matrix. *Journal of Occupational and Environmental Medicine* 54:184-191.

Colt, J. S., L. Stallones, L. L. Cameron, M. Dosemeci, and S. H. Zahm. 2001. Proportionate mortality among U.S. migrant and seasonal farmworkers in twenty-four states. *American Journal of Industrial Medicine* 40:604-611.

Cordell, R., L. Pickering, F. W. Henderson, and J. Murph. 2004. Infectious diseases in child-care settings. *Emerging Infectious Diseases* 10(11):e9.

Costello, S., A. M. Neophytou, D. M. Brown, E. M. Noth, S. K.Hammond, M. R. Cullen, and E. A. Eisen. 2016. Incident ischemic heart disease after long-term occupational exposure to fine particulate matter: Accounting for 2 Forms of Survivor Bias. *American Journal of Epidemiology* 183(9): 861-868.

CPWR (Center for Construction Research and Training). 2013. *The Construction Chart Book*. Available online at http://www.cpwr.com/publications/construction-chart-book (accessed April 5, 2017).

CPWR. 2017. *Fatality map*. Available online at http://stopconstructionfalls.com/fatality-map/ (accessed April 4, 2017).

CSTE (Council of State and Territorial Epidemiologists). 2013a. *Counting work-related injuries and illnesses: Taking steps to close the gaps*. Available online at http://c.ymcdn.com/sites/www.cste.org/resource/resmgr/OccupationalHealth/SummaryRecClosGapsMtg2013.pdf (accessed June 22, 2017).

CSTE. 2013b. *National assessment of epidemiologic capacity, findings and recommendations*. Available online at http://www.cste2.org/2013eca/CSTEEpidemiologyCapacityAssessment2014-final2.pdf (accessed September 4, 2017).

CSTE. 2014. *Position Statement 14-OH-02. Inclusion of work information as data elements in CDC surveillance systems*. Available online at www.cste.org/?page=Position Statements (accessed November 28, 2017).

CSTE. 2015a. *Handicapped-accessible accelerator pedal redesigned based on fatality*. Available online at http://c.ymcdn.com/sites/www.cste.org/resource/resmgr/Occupational_Health_Success_Stories/KY_success_story-_handicappe.pdf (accessed July 18, 2017).

CSTE. 2015b. *Public health reporting and national notification for elevated blood lead levels*. Available online at http://c.ymcdn.com/sites/www.cste.org/resourceresmgr/2015PS/2015PSFinal/15-EH-01.pdf (accessed June 14, 2017).

CSTE. 2017a. *Occupational health indicators*. Available online at http://www.cste.org/members/group.aspx?id=106668 (accessed June 10, 2017).

CSTE. 2017b. *Surveillance/informatics: Reportable Condition Knowledge Management System*. Available online at http://www.cste.org/group/RCKMS (accessed May 8, 2017).

Davis, L., M. Standbury, B. Materna, and E. Simms. 2012. *Putting data to work for worker safety and health, success stories in the states*. Available online at https://c.ymcdn.com/sites/cste.site-ym.com/resource/resmgr/OccupationalHealth/OC_Health_Book_FINAL.pdf (accessed May 1, 2017).

Du Y., X. Xu, M. Chu, Y. Guo, and J. Wang. 2016. Air particulate matter and cardiovascular disease: The epidemiological, biomedical and clinical evidence. *Journal of Thoracic Disease* 8(1): E8-E19.

Dubrow, R., J. Sestito, N. Lalich, C. Burnett, and J. Salg. 1987. Death certificate-based occupational mortality surveillance in the United States. *American Journal of Industrial Medicine* 11:329-342.

Edwards, C. H., G. S. Tomba, and B. F. de Blasio. 2016. Influenza in workplaces: Transmission, workers' adherence to sick leave advice and European sick leave recommendations. *European Journal of Public Health* 26(3):478-485.

Fagan, K. M., and M. J. Hodgson. 2017. Under-recording of work-related injuries and illnesses: An OSHA priority. *Journal of Safety Research* 60:79-83.

Fang, S. C., A. Cassidy, and D. C. Christiani. 2010. A systematic review of occupational exposure to particulate matter and cardiovascular disease. *International Journal of Environmental Research and Public Health* 7(4):1773-1806.

Froines, J. R., S. Baron, D. H. Wegman, and S. O'Rourke. 1990. Characterization of the airborne concentrations of lead in U.S. industry. *American Journal of Industrial Medicine* 18(1):1-17.

GAO (U.S. Government Accountability Office). 2015. *Contingent workforce: Size, characteristics, earnings, and benefits*. Available online at http://www.gao.gov/assets/670/669766.pdf (accessed June 14, 2017).

GAO. 2016. *Workplace Safety and Health: Additional Data Needed to Address Continued Hazards in the Meat and Poultry Industry*. GAO-16-337. Washington, DC: U.S. Government Printing Office.

Grattan K., L. K. Davis, M. Fiore, E. Pechter, and J. Laing. 2017. Employer Compliance with OSHA Requirement to Report Amputations, Updated Findings from a Massachusetts Study. CSTE Annual Conference, June, Boise, ID.

Green Seal Institutional Cleaning Products. 2017. *Green seal*. Available online at www.greenseal.org/FindGreenSealProductsandServices.aspx?vid=ViewProductDetail&cid=16 (accessed June 15, 2017).

Groenewold, M. R., and S. L. Baron. 2013. The proportion of work-related emergency department visits not expected to be paid by workers' compensation: Implications for occupational health surveillance, research, policy, and health equity. *Health Services Research* 48(6 Pt 1):1939-1959.

Haddon, W. 1970. On the escape of tigers: An ecologic note. *American Journal of Public Health* 60(2):2229–2234.

Harris, R. 2016. Suicide in the workplace. *Monthly Labor Review*. Available online at https://www.bls.gov/opub/mlr/2016/article/suicide-in-the-workplace.htm (accessed May 5, 2017).

Henneberger, P. K., C. A. Redlich, D. B. Callahan, P. Harber, C. Lemiere, J. Martin, S. M. Tarlo, and O. Vandenplas. 2011. An official American Thoracic Society statement: Work-exacerbated asthma. *American Journal of Respiratory Critical Care Medicine* 184(3):368-378.

HHS (U.S. Department of Health and Human Services). 2017. *Asthma*. Available online at https://www.healthypeople.gov/2020/topics-objectives/topic/respiratory-diseases/objectives (accessed April 5, 2017).

Hnizdo, E., P. A. Sullivan, K. M. Bang, and G. Wagner. 2002. Association between chronic obstructive pulmonary disease and employment by industry and occupation in the U.S. population: A study of data from the Third National Health and Nutrition Examination Survey. *American Journal of Epidemiology* 156:738-746.

Huband, E. M., and P. Bobbit. 2013. Nonresponse bias in the survey of occupational injuries and illnesses. BLS Statistical Survey Papers. Available online at https://www.bls.gov/osmr/pdf/st130170.pdf (accessed November 8, 2017).

IOM (Institute of Medicine). 2001. *Tuberculosis in the Workplace*. Washington, DC: National Academy Press.

IOM. 2009. *Traumatic Injury Research at NIOSH: Reviews of Research Programs of the National Institute for Occupational Safety and Health*. Washington, DC: The National Academies Press.

IRS (Internal Revenue Service). 2016. *New requirements for 501(c)(3) hospitals under the Affordable Care Act*. Available online at https://www.irs.gov/charities-non-profits/charitable-organizations/new-requirements-for-501c3-hospitals-under-the-affordable-care-act (accessed June 14, 2017).

Jackson L. 2001. Non-fatal occupational injuries and illnesses treated in hospital emergency departments in the United States. *Injury Prevention* 7(Suppl I):i21-i26.

Keech, M., and P. Beardsworth. 2008. The impact of influenza on working days lost: A review of the literature. *Pharmacoeconomics* 26(11):911-924.

Kica, J., and K. D. Rosenman. 2012. Multi-source surveillance system for work-related burns. *Journal of Occupational and Environmental Medicine* 54(5):642-647.

Kica, J., and K. D. Rosenman. 2014. Multi-source surveillance system for work-related skull fractures in Michigan. *Journal of Safety Research* 51:49-56.

Klous, G., A. Huss, D. J. Heederik, and R. A. Coutinho. 2016. Human-livestock contacts and their relationship to transmission of zoonotic pathogens, a systematic review of literature. *One Health* 2:65-76.

Largo, T. W., and K. D. Rosenman. 2015. Surveillance of work-related amputations in Michigan using multiple data sources: Results for 2006-2012. *Occupational and Environmental Medicine* 72(3):171-176.

LaMontagne A.D., T. Keegel, D.A. Vallance, A. Ostry, and R. Wolfe. 2008. Job strain—Attributable depression in a sample of working Australians: Assessing the contribution to health inequalities. *BMC Public Health* 8(181).

Lavoue, J., M. Friesen, and I. Burstyn. 2013a. Workplace measurements by the U.S. Occupational Safety and Health Administration since 1979: Descriptive analysis and potential uses for exposure assessment. *Annals of Occupational Hygiene* 57(5):681-683.

Lavoue, J., M. Friesen, and I. Burstyn. 2013b. Workplace measurements by the U.S. Occupational Safety and Health Administration since 1979: Descriptive analysis and potential uses for exposure assessment. *Annals of Occupational Hygiene* 57(1):77-97.

Lefkowitz, D., E. Pechter, M. Lumia, A. Stephens, K. Fitzsimmons, L. Davis, J. Flattery, J. Weinberg, R. J. Harrison, M. J. Reilly, M. S. Filios, G. E. White, and K. D. Rosenman. 2015. Isocyanates and work-related asthma: Findings from California, Massachusetts, Michigan, and New Jersey, 1993-2008. *American Journal of Industrial Medicine* 58(11): 1138-1149.

Luckhaupt, S., and G. M. Calvert. 2008. Deaths due to blood-borne infections and their sequelae among health care workers. *American Journal of Industrial Medicine* 51(11): 812-824.

Luckhaupt, S. E., J. M. Dahlhamer, B. W. Ward, A. L. Sussell, M. H. Sweeney, J. P. Sestito, and G. M. Calvert. 2013a. Prevalence of dermatitis in the working population, United States, 2010 National Health Interview Survey. *American Journal of Industrial Medicine* 56(6):625-634.

Luckhaupt, S. E., M. A. Cohen, and G. M. Calvert. 2013b. Concordance between current job and usual job in occupational and industry groupings: Assessment of the 2010 National Health Interview Survey. *Journal of Occupational and Environmental Medicine* 55(9):1074-1090.

MA COSH (Massachusetts Coalition for Occupational Safety and Health). 2005. *Protecting workers and homeowners from wood floor-finishing hazards in Massachusetts*. Available online at http://www.masscosh.org/files/ProtectingFromFloorFinishingHazards.pdf (accessed April 4, 2017).

MA DPH (Massachusetts Department of Public Health). 2017. *Fatal occupational injuries in Massachusetts: 2008-2013*. Available online at http://www.mass.gov/eohhs/docs/dph/occupational-health/fatal-reports/work-related-fatalities-2008-2013.pdf (accessed June 15, 2017).

Marcum J. L., and D. Adams. 2017. Work-related musculoskeletal disorder surveillance using the Washington State workers' compensation system: Recent declines and patterns by industry, 1999-2013. *American Journal of Industrial Medicine* 60(5):457-471.

Marsh, S. M., A. A. Reichard, R. Bhandari, and T. R. Tonozzi. 2016. Using emergency department surveillance data to assess occupational injury and illness reporting by workers. *American Journal of Industrial Medicine* 59(8):600-610.

Masterson, E. A., M. H. Sweeney, J. A. Deddens, C. L. Themann, and D. K. Wall. 2014. Prevalence of workers with shifts in hearing by industry: A comparison of OSHA and NIOSH hearing shift criteria. *Journal of Occupational and Environmental Medicine* 56(4):446-455.

Mazurek, J. M., G. Syamial, J. M. Wood, S. A. Hendricks, and A. Weston. 2017. Malignant mesothelioma mortality—United States, 1999-2015. *Morbidity and Mortality Weekly Report* 66:214-218.

Michaels, D. 2016. *Year one of OSHA's severe injury reporting program: An impact evaluation*. Available online at https://www.osha.gov/injuryreport/2015.pdf (accessed April 4, 2017).

Michas, M. G., and C. U. Iacono. 2008. Overview of occupational medicine training among US family medicine residency programs *Family Medicine* 40(2):102-106.

MI DCH (Michigan Department of Community Health). 2013. *Thirteen indicators of the health of Michigan's workforce*. Available online at http://www.michigan.gov/documents/Michigan_Indicator_Report_revised_41206_156036_7.pdf (accessed July 18, 2017).

MI DHHS (Michigan Department of Health and Human Services). 2015. *Heavy metals surveillance in Michigan: Eighth annual report*. Available online at http://www.oem.msu.edu/userfiles/file/Annual%20Reports/HeavyMetals/2014HeavyMetalsAnnualReport.pdf (accessed June 14, 2017).

MI FACE (Michigan Fatality Assessment and Control Evaluation). 2013. *Methylene chloride causes death of three Michigan bathtub refinishers. Hazard alert*. Available online at http://www.oem.msu.edu/userfiles/BathtubRefinishingHA14.pdf (accessed April 4, 2017).

Molinari, N. A., I. R. Ortega-Sanchez, M. L. Messonnier, W. W. Thompson, P. M. Wortley, E. Weintraub, and C. B. Bridges. 2007. The annual impact of seasonal influenza in the U.S.: Measuring disease burden and costs. *Vaccine* 25(27):5086-5096.

MSU and MI DELEG (Michigan State University and the Michigan Department of Energy, Labor, and Economic Growth). 2009. *Annual report on work-related noise-induced hearing loss in Michigan*. Available online at http://www.oem.msu.edu/userfiles/file/Annual%20Reports/Hearing/08NIHL_Report.pdf (accessed June 12, 2017).

NASEM (National Academies of Sciences, Engineering, and Medicine). 2017. *Information Technology and the U.S. Workforce: Where Are We and Where Do We Go from Here?* Washington, DC: The National Academies Press.

National COSH (Council for Occupational Safety and Health). 2017. *U.S. worker fatality maps*. Available online at http://www.coshnetwork.org/fatality-database-maps (accessed April 4, 2017).

NCHS (National Center for Health Statistics). 2016. *About the National Health Interview Survey*. Available online at https://www.cdc.gov/nchs/nhis/about_nhis.htm (accessed June 14, 2017).

NCHS. 2017. *National Health Interview Survey 2018 questionnaire redesign: Proposed design*. Available online at https://www.cdc.gov/nchs/nhis/2018_quest_redesign.htm (accessed November 27, 2017).

New York State Department of Health. 2016. *Heavy metals surveillance: New York State heavy metals registry*. Available online at https://www.health.ny.gov/environmental/workplace/heavy_metals_registry (accessed June 14, 2017).

Niedhammer I., H. Sultan-Taieb, J. F. Chastang, G. Vermeylen, and A. Parent-Thirion. 2014. Fractions of cardiovascular diseases and mental disorders attributable to psychosocial work factors in 31 countries in Europe. *International Archives of Occupational and Environmental Health* 87(4):403-411.

NIOSH (National Institute for Occupational Safety and Health). 1974. *National Occupational Hazard Survey. Volume I: Survey Manual*. Cincinnati, OH: NIOSH.

NIOSH. 1997. *Musculoskeletal Disorders and Workplace Factors - A Critical Review of Epidemiologic Evidence for Work-Related Musculoskeletal Disorders of the Neck, Upper Extremity, and Low Back*. Cincinnati, OH: NIOSH. Available online at https://www.cdc.gov/niosh/docs/97-141/pdfs/97-141.pdf (accessed November 8, 2017).

NIOSH. 2012. *A Story of Impact: NIOSH Research Cited in Recommendations for Improving Commercial Fishing*. NIOSH Publication No. 2012-129 March 2012. Available online at https://www.cdc.gov/niosh/docs/2012-129/pdfs/2012-129.pdf (accessed May 5, 2017).

NIOSH. 2013. *2010 occupational health supplement*. Available online at https://www.cdc.gov/niosh/topics/nhis/healthcareocc/hcocctables.html (accessed December 1, 2017).

NIOSH. 2014a. *Impact: PFD manufacturer adopts NIOSH research into product development process*. Available online at https://www.cdc.gov/niosh/docs/2015-119/pdfs/2015-119.pdf (accessed May 6, 2017).

NIOSH. 2014b. *NIOSH research projects*. Available online at https://www.cdc.gov/niosh/programs/surv/projects2.html#ohd (accessed June 14, 2017).

NIOSH. 2015a. *Washington occupational injury and illness surveillance program—Annual report*. Available online at https://www.cdc.gov/niosh/oep/pdfs/annualreports/washington-occupational-injury-and-illness surveillance-program.pdf (accessed November 17, 2017).

NIOSH. 2015b. *Health hazard evaluations*. Available online at https://www.cdc.gov/niosh/hhe/resources.html (accessed May 8, 2017).

NIOSH. 2016a. *Definitions of traumatic injuries and MSDs*. Available online at https://www.cdc.gov/niosh/programs/msd/risks.html (accessed June 13, 2017).

NIOSH. 2016b. *Coal Workers' Health Surveillance Program (CWHSP) data query system*. Available online at https://webappa.cdc.gov/ords/cwhsp-database.html (accessed December 1, 2017).

NIOSH. 2016c. *National agriculture, forestry and fishing agenda*. Available online at https://www.cdc.gov/niosh/nora/comment/agendas/agforfish/default.html (accessed June 13, 2017).

NIOSH. 2016d. *National Health Interview Survey: Occupational health supplement*. Available online at https://www.cdc.gov/niosh/topics/nhis/ (accessed April 5, 2017).

NIOSH. 2016e. *Tuberculosis*. Available online at https://www.cdc.gov/niosh/topics/tb/ (accessed May 8, 2017).

NIOSH. 2016f. *Adult blood lead epidemiology and surveillance*. Available online at http://www.cdc.gov/niosh/topics/ables/description.html (accessed November 21, 2016).

NIOSH. 2016g. *NIOSH musculoskeletal disorders prevention programs*. May. Available online at https://stacks.cdc.gov/view/cdc/39919 (accessed December 21, 2017).

NIOSH. 2017a. *Fatality Assessment and Control Evaluation (FACE) program*. Available online at https://www.cdc.gov/niosh/face/default.html (accessed April 4, 2017).

NIOSH. 2017b. *Firefighter fatality investigation and prevention*. Available online at https://www.cdc.gov/niosh/fire/default.html (accessed January 29, 2017).

NIOSH. 2017c. *NIOSH and partners work to prevent worker deaths from exposures to hydrocarbon gases and vapors at oil and gas wellsites*. Available online at https://www.cdc.gov/niosh/docs/2017-110 (accessed May 8, 2017).

NIOSH. 2017d. *Welcome to eWORLD*. Available online at https://wwwn.cdc.gov/eworld (accessed June 14, 2017).

NIOSH. 2017e. *Coal workers' health surveillance program*. Available online at https://www.cdc.gov/niosh/topics/cwhsp (accessed April 5, 2017).

NIOSH. 2017f. *Bloodborne infectious diseases: HIV/AIDS, hepatitis B, hepatitis C*. Available online at https://www.cdc.gov/niosh/topics/bbp (accessed May 8, 2017).

NIOSH. 2017g. *Health and safety practices survey of healthcare workers*. Available online at https://www.cdc.gov/niosh/topics/healthcarehsps/aboutstudy.html (accessed June 14, 2017).

NIOSH. 2017h. *National Occupational Exposure Survey (NOES)*. Available online at https://www.cdc.gov/noes/default.html (accessed November 8, 2017).

NRC (National Research Council). 1987. *Counting Injuries and Illnesses in the Workplace: Proposals for a Better System*. Washington, DC: National Academy Press.

NRC. 2001. *Musculoskeletal Disorders and the Workplace: Low Back and Upper Extremities*. Washington, DC: National Academy Press.

O'Leary P., L. I. Boden, S. A. Seabury, A. Ozonoff, and E. Scherer. 2012. Workplace injuries and the take-up of Social Security benefits. *Social Security Bulletin* 72(3).

OMB (Office of Management and Budget). 2007. Implementation Guidance for Title V of the E-Government Act, Confidential Information Protection and Statistical Efficiency Act of 2002 (CIPSEA). Notice of Decision. 72 *Federal Register* 33362-33377, June 15.

OSHA (Occupational Safety and Health Administration). 2001. Occupational Injury and Illness Recording and Reporting Requirements. Final Rule. 66 *Federal Register* 5921-5923, January 19.

OSHA. 2007. *Hispanic outreach*. Available online at www.osha.gov/OshDoc/data_Hispanic/hispanic_outreach.pdf (accessed June 12, 2017).

OSHA. 2013. *Methylene chloride hazards for bathtub refinishers*. Available online at https://www.osha.gov/dts/hazardalerts/methylene_chloride_hazard_alert.pdf (accessed May 8, 2017).

OSHA. 2014. Temporary Worker Initiative. *TWI Bulletin* 1. Available online at https://www.osha.gov/temp_workers/OSHA_TWI_Bulletin.pdf (accessed April 4, 2017).

OSHA. 2015. *OSHA fact sheet: Preventing cuts and amputations from food slicers and meat grinders*. Available online at https://www.osha.gov/Publications/OSHA3794.pdf (accessed June 13, 2017).

OSHA, 2016a. Improve Tracking of Workplace Injuries and Illnesses. Final Rule. 81 *Federal Register* 29637-29639, May 12.

OSHA. 2016b. *Memorandum for regional administrators and state designees from Thomas Galassi, director, Directorate of Enforcement Programs, "Revised Interim Enforcement Procedures for Reporting Requirements under 29 CFR 1904.39," March 4*. Available online at https://www.osha.gov/dep/enforcement/Interm_Enforcement_Procedures.html (accessed April 5, 2017).

OSHA. 2017a. *National safety stand-down to prevent falls in construction*. Available online at https://www.osha.gov/StopFallsStandDown (accessed April 4, 2017).

OSHA. 2017b. *Fatality and catastrophe investigation summaries*. Available online at www.osha.gov/pls/imis/accidentsearch.html (accessed April 4, 2017).

OSHA. 2017c. *Severe injury reports*. Available online at https://www.osha.gov/severeinjury/index.html (accessed May 8, 2017).

Pegula, S. M. 2013. An analysis of fatal occupational injuries at road construction sites, 2003–2010. *Monthly Labor Review* November. Available online at https://www.bls.gov/opub/mlr/2013/article/an-analysis-of-fatal-occupational-injuries-at-road-construction-sites-2003-2010.htm (accessed April 4, 2017).

Pegula, S. M., and A. Measure. 2016. *Web scraping and a timely repository for fatality data*. Presentation to the Data Users Advisory Committee, Bureau of Labor Statistics. November 10.

PHAB (Public Health Accreditation Board). 2015. *Guide to national public health department initial accreditation*. Available online at http://www.phaboard.org/wp-content/uploads/Guide-to-Accreditation-final_LR2.pdf (accessed June 15, 2017).

PHII (Public Health Informatics Institute). 2016. *Next generation electronic death registration system: Supporting improved quality and timeliness of vital records data.* Available online at https://www.phii.org/sites/www.phii.org/files/resource/pdfs/20160316%20Next%20Generation%20EDRS%20Report%20FINAL%20WEBSITE.pdf (accessed June 13, 2017).

Raherison, C., and P. O. Girodet. 2009. Epidemiology of COPD. *European Respiratory Review* 18:213-221.

Rappin, C. L., S. E. Wuellner., and D. K. Bonauto. 2016. Employer reasons for failing to report eligible workers compensation claims in the BLS Survey of Occupational Injuries and Illnesses. *American Journal of Industrial Medicine* 59(5):343-356.

Reves, R. R., and L. K. Pickering. 1992. Impact of child day care on infectious diseases in adults. *Infectious Disease Clinics of North America* 6(1):239-250.

Reville, R. T. and R. F. Schoeni. 2004. The fraction of disability caused at work. *Social Security Bulletin* 65(4). Available online at https://www.ssa.gov/policy/docs/ssb/v65n4/v65n4p31.html (accessed November 8, 2017).

Robinson, C. F., J. T. Walker, M. H. Sweeney, R. Shen, G. M. Calvert, P. K. Schumacher, J. Ju, and S. Nowlin. 2015. Overview of the National Occupational Mortality Surveillance (NOMS) system: Leukemia and acute myocardial infarction risk by industry and occupation in 30 U.S. states 1985-1999, 2003-2004, and 2007. *American Journal of Industrial Medicine* 58(2):123-137.

Rosenman, K. D. 2016. OSHA, well past its infancy, but still learning how to count injuries and illnesses. *American Journal of Industrial Medicine* 59(8):595-599.

Rosenman, K. D., M. J. Reilly, D. P. Schill, D. Valiante, J. Flattery, R. Harrison, F. Reinisch, E. Pechter, L. Davis, C. M. Tumpowsky, and M. Fillios. 2003. Cleaning products and work-related asthma. *Journal of Occupational and Environmental Medicine* 45(5):556-563.

Rosenman, K. D., A. Kalush, M. J. Reilly, J. C. Gardiner, M. Reeves, and Z. Luo. 2006. How much work-related injury and illness is missed by the current national surveillance system? *Journal of Occupational and Environmental Medicine* 48(4):357-365.

Ruser, J. W. 1998. Denominator choice in the calculation of workplace fatality rates. *American Journal of Industrial Medicine* 33(2):151-156.

Ruser, J. W. 2008. Examining evidence on whether BLS undercounts workplace injuries and illnesses, *Monthly Labor Review*, August 2008. Available online at https://www.bls.gov/opub/mlr/2008/08/art2full.pdf (accessed November 17, 2017).

Rutstein, D. D., R. J. Mullan, T. M. Frazier, W. E. Halperin, J. M. Melius, and J. P. Sestito. 1983. Sentinel health events (occupational): A basis for physician recognition and public health surveillance. *American Journal of Public Health* 73(9):1054-1062.

Sabbath, E. L., L. I. Boden, J. A. Williams, D. Hashimoto, K. Hopcia, and G. Sorensen. 2017. Obscured by administrative data? Racial disparities in occupational injury. *Scandinavian Journal of Work, Environment, and Health* 43(2):155-162.

Sarazin, P., I. Burstyn, L. Kincl, and J. Lavoue. 2016. Trends in OSHA compliance monitoring data 1979-2011: Statistical modeling of ancillary information across 77 chemicals. *Annals of Occupational Hygiene* 60(4):432-452.

Savitz, D. A., D. P. Loomis, and C. K. Tse. 1998. Electrical occupations and neurodegenerative disease: Analysis of U.S. mortality data. *Archives of Environmental Health* 53:71-74.

SBU (Swedish Agency for Health Technology Assessment and Assessment of Social Services). 2017. *Occupational health and safety–Chemical exposure.* Available online at http://www.sbu.se/261e (accessed May 17, 2017).

Schleiff, P. L., J. M. Mazurek, M. J. Reilly, K. D. Rosenman, M. B. Yoder, M. E. Lumia, and K. Worthington. 2016. Surveillance for silicosis—Michigan and New Jersey, 2003–2011. *Morbidity and Mortality Weekly Report* 63(55):73-78.

Schnorr, T. 2016. Sponsor panel for follow-up questions and answers. Presentation to the National Academies Committee on Developing a Smarter National Surveillance System for Occupational Safety and Health in the 21st Century, September 21.

Sears, J. M., and S. M. Bowman. 2016. State trauma registries as a resource for occupational injury surveillance and research: Lessons from Washington State, 1998-2009. *Public Health Reports* 131(6):791-799.

Simpson, H. 2016. Sponsor panel for follow-up questions and answers. Presentation to the National Academies Committee on Developing a Smarter National Surveillance System for Occupational Safety and Health in the 21st Century, September 21.

Sincavage, J. R. 2005. Fatal occupational injuries among Asian workers. *Monthly Labor Review* October:49-55. Available online at https://www.bls.gov/opub/mlr/2005/10/art6full.pdf (accessed April 4, 2017).

Smith, G. S., H. M. Wellman, G. S. Sorock, M. Warner, T. K. Courtney, G. S. Pransky, and L. A. Fingerhut. 2005. Injuries at work in the U.S. adult population: Contributions to the total injury burden. *American Journal of Public Health* 95(7):1213-1219.

Socias, C. M., C. K. Chaumont Menendez, J. W. Collins, and P. Simeonov. 2014. Occupational ladder fall injuries—United States, 2011. *Morbidity and Mortality Weekly Report* 63(16):341-346.

Spieler, E. A., and J. F. Burton, Jr. 2012. The lack of correspondence between work-related disability and receipt of workers' compensation benefits. *American Journal of Industrial Medicine* 55(6):487-505.

Spieler, E. A., and G. R. Wagner. 2014. Counting matters: Implications of undercounting in the BLS Survey of Occupational Injuries and Illnesses. *American Journal of Industrial Medicine* 57(10):1077-1084.

Stanbury, M., A. P. Rafferty, and K. D. Rosenman. 2008. Prevalence of hearing loss and work-related noise induced hearing loss in Michigan. *Journal of Occupational and Environmental Medicine* 50(1):72-79.

Steenland, K., C. Burnett, N. Lalich, E. Ward, and J. Hurrell. 2003. Dying for work: The magnitude of U.S. mortality from selected causes of deaths associated with occupation. *American Journal of Industrial Medicine* 43(5):461-482.

Tak, S., R. R. Davis, and G. M. Calvert. 2009. Exposure to hazardous workplace noise and use of hearing protection devices among U.S. workers—NHANES, 1999-2004. *American Journal of Industrial Medicine* 52(5):358-371.

Thacker, S. B., J. R. Qualters, and L. M. Lee. 2012. Public health surveillance in the United States: Evolution and challenges. *Morbidity and Mortality Weekly Report* 61(3):3-9.

Theorell T., A. Hammarström, G. Aronsson, L. T. Bendz, T. Grape, C. Hogstedt, I. Marteinsdottir, I. Skoog, and C. Hall. 2015. A systematic review including meta-analysis of work environment and depressive symptoms. *BMC Public Health* 15:738.

Thomsen, C., J. McClain, K. Rosenman, and L. Davis. 2007. Indicators for occupational health surveillance. *Morbidity and Mortality Weekly Report* 56(RR-1):1-7.

Tonozzi, T. R., S. M. Marsh, A. A. Reichard, and R. Bhandari. 2016. Reported work-related injuries and illnesses among Hispanic workers: Results from an emergency department surveillance system follow-back survey. *American Journal of Industrial Medicine* 59(8):622-630.

Torén, K., and P. D. Blanc. 2009. Asthma caused by occupational exposures is common—a systematic analysis of estimates of the population-attributable fraction. *BMC Pulmonary Medicine* 9:7, doi: 10.1186/1471-2466-9-7.

Weber, D. J., and W. A. Rutala. 2016. Occupational health update: Focus on preventing the acquisition of infections with pre-exposure prophylaxis and postexposure prophylaxis. *Infectious Disease Clinics of North America* 30(3):729-757.

Weil, D. 2014. *The Fissured Workplace: Why Work Became So Bad for So Many and What Can Be Done to Improve It.* Cambridge, MA: Harvard University Press.

Weil, D. 2017. How to make employment fair in an age of contracting and temp work. *Harvard Business Review*, March 24. Available online at https://hbr.org/2017/03/making-employment-a-fair-deal-in-the-age-of-contracting-subcontracting-and-temp-work (accessed June 15, 2017).

Wiatrowski, W. J. 2005. Fatalities in the ornamental shrub and tree services industry. *Compensation and Working Condition*, July 25. Available online at https://www.bls.gov/opub/mlr/cwc/fatalities-in-the-ornamental-shrub-and-tree-services-industry.pdf (accessed April 4, 2017).

Wiatrowski, W. J. 2014. The BLS Survey of Occupational Injuries and Illnesses: A primer. *American Journal of Industrial Medicine* 57(10):1085-1089.

Wilken, J. A., G. Sondermeyer, D. Shusterman, J. McNary, D. J. Vugia, A. McDowell, P. Borenstein, D. Gilliss, B. Ancock, J. Prudhomme, D. Gold, G. C. Windham, L. Lee, and B. L. Materna. 2015. Coccidioidomycosis among workers constructing solar power farms, California, USA, 2011-2014. *Emerging Infectious Diseases* 21(11):1997-2005.

Windau, J. A. 1998. Worker fatalities from being caught in machinery. *Compensation and Working Condition* Winter:35-38. Available online at https://www.bls.gov/opub/mlr/cwc/worker-fatalities-from-being-caught-in-machinery.pdf (accessed June 12, 2017).

Windau, J., K. D. Rosenman, H. Anderson, L. Hanrahan, L. Rudolph, M. Stanbury, and A. Stark. 1991. The identification of occupational lung disease from hospital discharge data. *Journal of Occupational Medicine* 33(10):1061-1066.

Wuellner, S. E., and D. K. Bonauto. 2014. Exploring the relationship between employer recordkeeping and underreporting in the BLS Survey of Occupational Injury and Illness. *American Journal of Industrial Medicine* 57(10):1133-1143.

Wuellner, S., and P. Phipps. 2016. Identifying patterns in employer reporting errors in the BLS Survey of Occupational Injuries and Illnesses. Pp. 3322-3335 in *JSM Proceedings, Statistical Computing Section*. Alexandria, VA: American Statistical Association. Available online at https://www.bls.gov/osmr/pdf/st160200.pdf (accessed April 4, 2017).

Wuellner, S. E., D. A. Adams, and D. K. Bonauto. 2016. Unreported workers' compensation claims to the BLS Survey of Occupational Injuries and Illnesses: Establishment factors. *American Journal of Industrial Medicine* 59(4):274-279.

Wuellner, S. E., D. A. Adams, and D. K. Bonauto. 2017. Workers' compensation claims not reported in the Survey of Occupational Injury and Illnesses: Injury and claim characteristics. *American Journal of Industrial Medicine* 60(3):264-275.

WY DWS (Wyoming Department of Workforce Services). 2016. *Wyoming state occupational epidemiologist releases most comprehensive report on workplace fatalities since program's inception.* Available online at http://www.wyomingworkforce.org/news/2016-10-25a (accessed May 8, 2017).

# 5

# International Approaches to Occupational Health Surveillance

An exploration of the occupational safety and health surveillance systems of other nations is useful in informing the design and development of occupational health and safety efforts in the United States. Although no other national surveillance system is directly comparable, primarily due to the structure of the workers' compensation system in the United States, an overview of the characteristics of some of these systems informs possibilities for the evolution of the U.S. system. This chapter follows the organization of the previous chapter and is structured according to four general categories: occupational fatalities, occupational injuries and acute illnesses, occupational disease, and hazards and exposures. The work of Tedone (2017) was a useful resource in developing this chapter.

## OCCUPATIONAL FATALITIES

The Census of Fatal Occupational Injuries (CFOI) system (described in Chapter 4) for identifying all occupational injury fatalities in the United States is different from and likely more comprehensive than those in other industrialized nations. For example, the major difference between CFOI and fatal occupational injuries published by countries contributing data to Eurostat is that the CFOI is an active surveillance system and Eurostat relies on passively reported fatalities (Benavides et al., 2003). As a consequence, it is believed that the CFOI system provides a more complete assessment of fatalities from work. Australia maintains a Work-Related Traumatic Injury Fatalities system that, like CFOI, collects reports of fatal injuries from a variety of sources (Safe Work Australia, 2017a). The Australian system also includes as work fatalities those that occur when commuting to or from work.

## OCCUPATIONAL INJURIES

Occupational injury surveillance generally includes both traumatic injuries and acute occupational disease. Surveillance of most chronic occupational disease is considered below. Musculoskeletal disorders are often managed as injuries although they may result from repetitive trauma over time.

Occupational injury surveillance is managed in a variety of different ways by different countries. Differences include source of data, types of injuries covered, and work population exclusions. In general, there are two types of approaches and often countries will use a combination of these: compensation-based and noncompensation schemes. The scope and coverage of each national scheme varies somewhat depending on the source of data used for surveillance: employer records (establishment surveys), employee interviews (household or labor force surveys), administrative compensation data (workers' compensation), and regulatory reporting (mandated reports to enforcement authorities). As illustrated below, those countries with mandatory accident insurance generally use records of compensation as the primary source of occupational injury data. By contrast, the United States relies more on regulatory reporting, databases, and statistical methodologies such as surveys (Walters, 2007).

### Workers' Compensation

Many developed nations have some type of national compensation scheme often integrated with their national health systems. Although the specific coverages vary by country the compensated injuries provide counts of injuries that qualify. These systems are passive but have national scope and, in that way, are quite different from the state-based systems in the United States. Canada is the exception, having a provincial-based system where accepted workers' compensation claims may differ among provinces. Nonetheless, accepted claims are forwarded to a single source using a common format concerning standard data elements for collating and reporting on compensated injuries (AWCBC, 2017). The European Union (EU) members have agreed to a standard for reporting occupational injuries that account for a minimum of 3 days away from work (Eurostat, 2017a). Australia nonfatal injury reports include only cases that have been accepted for workers' compensation and have a minimum of 1 week away from work (Safe Work Australia, 2017b). New Zealand occupational injury reports include data from all accepted claims along with a subgroup of those claims characterized by disability or time away from work (WorkSafe New Zealand, 2017). Japan collects compensated case information for those with 4 or more days away from work (JISHA, 2017).

### Establishment Surveys

The committee identified several surveys worldwide that, similar to the Survey of Occupational Injuries and Illnesses, use a statistical sample of employers annually to collect records of occupational injuries or acute diseases. Japan conducts two different surveys, one for general industry and one for the construction industry. These surveys are limited to establishments of 10 or more employees and in general published data are provided for establishments with 100 or more employees (Japan Ministry of Health, Labour, and Welfare, 2017). The response rate for this survey is not reported. Although the European Union has no common requirement for an establishment survey, some countries undertake such an effort. The Netherlands National Organisation for Applied Scientific Research, for example, undertakes a periodic survey of employers (Employers' Labour Survey) but the response rate has been low. The committee did not identify work establishment surveys in Canada, Australia, and New Zealand that collect occupational injury statistics.

### Labor Force Surveys

Many countries supplement occupational injury information obtained from workers' compensation systems with reports from a household survey. The prime example of this approach is represented by the United Kingdom's (UK's) Labour Force Survey (LFS). The LFS was developed as a condition of membership in the European Union with a purpose ". . . to provide information on the UK labour market which can then be used to develop, manage, evaluate and report on labour market policies"( HSE, 2017a). The Self-reported Work-related Illness (SWI) and Workplace Injury modules were introduced in 1990 and these have become annual since 1993. The modules include a core set of questions for those aged 16 or over who are currently employed or who have been employed in the previous year. The LFS is considered the preferred source (see Table 5-1) for occupational injury data in the UK because "data from the LFS represents the views of workers, providing a more complete view than RIDDOR, as it is not subject to changes in legislation or operational activity, nor subject to substantial levels of under reporting" (HSE, 2017b). These characteristics of a population survey are an important input for estimating burden of occupational injury.

Australia, in addition to its compensation-based primary source of work-related injuries (see above), has supplemented its labor force survey

**TABLE 5-1** UK Preferred Sources for Injuries and Illnesses

*These ratings provide a general indication of the source's strength for most purposes. Please note that there may be a preferred source for a particular item, but another source may be more appropriate for a particular data breakdown of the item.*

Strength of Source: = strong, = weak

| Other Source(s)Nature of Harm | Preferred Source | Other Source(s) | | |
|---|---|---|---|---|
| Injuries | LFS | RIDDOR | | |
| *Common conditions arising in a wide range of occupational settings* | | | | |
| Musculoskeletal disorders | LFS | THOR-GP | | |
| Stress | LFS | THOR-GP | IIDB | |
| *Common conditions arising in a limited range of occupational settings* | | | | |
| Asthma | SWORD | THOR-GP | IIDB | LFS |
| COPD | AF | IIDB | SWORD | |
| Cancer | AF | IIDB | SWORD | EPIDERM |
| Noise induced hearing loss | AF | LFS | IIDB | |
| Skin disorders / dermatitis | EPIDERM | THOR-GP | LFS | IIDB |
| *Specific or rare conditions arising in a limited range of occupational settings* | | | | |
| Asbestos-related lung cancer | AF | SWORD | IIDB | |
| Diffuse pleural thickening | IIDB | SWORD | | |
| Hand-arm vibration | AF | IIDB | | |
| Mesothelioma | DC | IIDB | | |
| Pneumoconiosis & silicosis | IIDB | SWORD | DC | |
| Other respiratory disease | SWORD | IIDB | DC | |

Acronym Key

- **RIDDOR** – Reporting of Injuries, Diseases and Dangerous Occurrences Regulations
- **LFS** – Labour Force Survey
- **THOR** – the Health and Occupation Reporting network
- **SWORD** – Surveillance of Work-related and Occupational Respiratory Disease specialist surveillance scheme
- **EPIDERM** – Skin specialist surveillance scheme
- **IIDB** – Industrial Disablement Benefit
- **DC** – Death certificate
- **AF** – Attributable fraction – This is the proportion of the total number of cases of the disease that are caused by occupational exposure. The AF is either measured directly from a bespoke study or estimated by combining and comparing relevant epidemiological studies from around the world.

SOURCE: HSE, 2017b.

with questions on occupational injury and acute illnesses periodically (2005-2006, 2009-2010, and 2013-2014) (Australian Bureau of Statistics, 2013). Some limited information is also collected as part of the Survey of Employment Arrangements and Superannuation and the National Health Survey. The European Union has an extensive system of labor force surveys that are based on household sample and provides results quarterly on aspects of labor force participation (Eurostat, 2017b). These surveys are carried out by each member country according to a common format and structure then centrally managed by Eurostat. These surveys have been supplemented with modules about work injuries in 2007 and 2013. In addition, the European Foundation for the Improvement of Living and Working Conditions conducts European Working Conditions Surveys (see discussion below).

### Mandated Reporting

The United Kingdom uses a reporting program, the Reporting of Injuries, Diseases and Dangerous Occurrences Regulations (RIDDOR), which obligates the employer or self-employed individual to report fatalities, injuries, or illnesses of specific types, and incidents that could lead to serious harm. The RIDDOR system serves as a secondary source of information for estimates of burden but primarily provides information to enhance prevention and to inform actions by the Health and Safety Executive (HSE) to address risk reduction in workplaces (HSE, 2017c). New Zealand supplements its compensation data with reports of serious injury outcome indicators that are derived according to public health categories of national importance (assaults, work-related injuries, suicide and intentional self-harm, falls, motor-vehicle injuries, and drowning or near drowning) (Statistics New Zealand, 2014). Serious injury outcome indicator events are collected from national mortality and hospital discharge systems. Among the international occupational and safety health surveillance systems examined by the committee, these are the only two national noncompensation-based mandatory systems.

### Injuries that Occur During Commuting to Work

Injuries that occur during commuting to work is a category of work-related injury that is not dealt with consistently by the different national occupational injury systems. Apart from England and Ireland, European nations include road traffic injuries in their occupational statistics. In the United States, road traffic injuries are not included in occupational safety and health (OSH) surveillance systems if the fatality occurred while commuting or as a bystander. Many European nations, however, count commuting injuries as one category of occupational injuries. A summary of experience from Belgium, Finland, France, and Germany reported that between 10 and 15 percent of occupational injuries occur as a result of commuting and approximately 45 percent of fatalities were related to commuting (Munich Re Group, 2004). There is some evidence that commuting distances to work are increasing (Jarosz and Cortes, 2014; Kneebone and Holmes, 2015), which increases the probability that commuting collisions will occur. Furthermore, one study from France found differential risk of commuting injuries among certain work groups. Women employed in health and community service professions had a threefold excess risk of a commuting accident and male sales workers had a sixfold risk (Hours et al., 2011). Consideration needs to be given to include commuting fatalities and injuries as work related.

## OCCUPATIONAL DISEASE

Surveillance of occupational disease is challenging for a number of reasons that apply in many countries. An underlying challenge is the lack of knowledge by general practitioners and medical specialists about the role of work as a primary factor or as one of several factors that can cause or exacerbate disease. This problem is compounded by whether the physician reports the occupational disease, how physicians understand the compensation system, and the efficiency of the reporting system. Furthermore, patient care-seeking behavior may or may not be coupled with whether the patient reports the possibility of an association of symptoms with work. There may also be additional factors that can hamper occupational disease surveillance, including employer pressures on physicians and workers not to report the relationship of a disease with work, workers' fears of the consequences of reporting, and undeclared or informal work.

### Schedules of Occupational Diseases

The challenges noted above have led many nations to rely on schedules (lists) of occupational diseases such as those developed by the European Union or the International Labour Organization (ILO) as the primary means of identifying the work relationship of a disease (ILO, 2010; European Agency for Safety and Health at Work, 2017). Many of the diseases are included because they are associated with specific work-related exposures known to cause the disease. Schedules are established on a national basis and primarily serve to identify conditions that are eligible for workers' compensation (European Commission, 2013). There is variation among countries in the list of occupational diseases and in the presumption of work-related origin. The lists of occupational diseases developed by both the EU and the ILO are similar, with the EU making a distinction between recognized and suspected diseases and ILO combining both in one list. Since 2003, five EU members have adopted the full EU list while the rest vary in what is considered an occupational disease. The data for the reporting of occupational diseases is primarily derived from the compensation or social security system while a few use an independent system. Underreporting was assessed in a EUROGIP study (2015) where the experience in five countries that represented the diverse insurance models were compared. A fourfold difference was seen between the highest and lowest claim rates and, although in different relative proportion, the same diseases led reporting in all countries: musculoskeletal disorders, hearing loss, skin diseases, and cancer.

Canada collects evidence of occupational diseases through its compensation system. Although there is a reporting requirement for employers to report occupational diseases that they become aware of, this system is likely much less useful than the reports from the compensation programs. A summary assessment of occupational diseases suggests that "[v]irtually every board in Canada would accept any of the 29 groups of occupational diseases recognized by the International Labour Organization as being work-related, but acceptance of less well-established conditions, such as suspected but not proven causes of cancer, would vary among the boards" (Guidotti, 2013). Japan appears to collect information on occupational disease through their compensation system (physician reporting is not mandated) and utilizes a form of a list for eligible conditions (Takahashi and Ishii, 2014). New Zealand uses a list of occupational diseases coupled with a voluntary notification system (Armstrong and Bunn, 2012). Reviews of occupational disease reporting in New Zealand suggested that only the compensation statistics were useful and those underestimated the true count (Driscoll et al., 2004; Pearce et al., 2005).

## Multisystem Approaches

The United Kingdom has the most comprehensive approach to occupational disease surveillance that includes required reporting of one of eight conditions included in a published list, reports of cancer or acute infectious diseases meeting minimal criteria, and reports from a voluntary physician reporting system called The Health and Occupation Research (THOR) network (Carder et al., 2017). The occupational diseases required to be reported by employers and the self-employed are carpal tunnel syndrome, severe cramp of the hand or forearm, occupational dermatitis, hand-arm vibration syndrome, occupational asthma, tendonitis or tenosynovitis of the hand or forearm, any occupational cancer, and any disease attributed to an occupational exposure to a biological agent. Cancer cases are reported if there is an established link and the worker was exposed to the agent(s). Illness due to a biological agent is reportable if it occurs in association with an accident that created likely exposure or an infection reasonably associated with a work exposure. Guidance is provided for each of these categories (HSE, 2017d).

The THOR physician reporting approach is innovative and includes reports of any occupational disease by general practitioners and specific occupational conditions by specialist physicians. THOR-GP began in 2005 and utilizes more than 200 general practitioners as the source of reports of occupational disease. These reporters have received postgraduate education in occupational medicine and practice and are distributed throughout the United Kingdom. Currently these physicians are asked to report conditions that "more likely than not" are work related. They do so for one randomly selected month each year. Audits of the system suggest that underreporting is common due to the small number of reporters (approximately 1 percent of general practitioners in the United Kingdom), only captures conditions seen in practices of general practitioners, and depends on patient care-seeking and reporting behavior. However, efforts have been made to estimate burden from these reports, which help focus revisions to the system (Hussey, 2016).

THOR specialist reporting schemes have been introduced for several conditions and two appear to be particularly successful: Surveillance of Work-related Occupational Respiratory Disease (based on reports from respiratory disease specialists) and the skin specialist surveillance scheme (EPIDERM) (based on reports from consultant dermatologists). Both have proved useful in characterizing the distribution and burden of the two conditions (HSE, 2016a,b). Both schemes suffer the same limitations as THOR-GP but the reporting appears more consistent and has led to useful investigations (Stocks et al., 2011; Warburton et al., 2015). Nonetheless, the HSE considers these two systems superior to the Labour Force Survey for occupational asthma and skin disorders (Table 5-1). Other efforts, no longer operating, have focused on occupational physicians and rheumatologists.

The THOR system has notably been explored for surveillance of mental health with some success. From 1999-2009, an effort was organized for reporting by consultant psychiatrists to the Surveillance of Occupational Stress and Mental Illness program. Over 3,500 case reports were used by the HSE to identify areas and occupational categories at greatest risk from occupational mental ill-health. During the same period and continuing today, the general practitioners participating in THOR-GP also reported cases of mental ill-health among their patients. Examination of the mental health burden of work suggests that these reporting schemes prove useful in surveillance of mental health at work (Agius and Turner, 2004; Hussey et al., 2013; Zhou et al., 2017).

## WORK HAZARDS OR EXPOSURES

Most occupational health surveillance focuses on outcomes rather than exposures or hazards. While this proves less of a problem for occupational injuries, which are acute events, its importance is evident when considering work-related diseases. Many occupational diseases result long after the exposures are initiated or even well after employment with the relevant exposure ceases. Therefore, currently diagnosed occupational diseases often reflect exposures that occurred in the past. Formal study of these conditions may improve understanding of the etiology but exposure (and therefore risk) continues. When the exposure-disease connection is known, tracking these exposures and acting to reduce or eliminate them is a sound public health approach (see Chapter 6).

### Finnish Job Exposure Matrix

The Finnish Job Exposure Matrix (FINJEM) provides an example of a surveillance system focused on assessing the extent of worker exposure to chemical and physical agents. Estimates from FINJEM of the extent of exposures experienced by Finnish workers are used as the national source for quantitative risk assessments (Kauppinen et al., 2014). Updated exposure estimates are provided every 3 years for most agents (Kauppinen et al., 2014). Cross-sectional data, data on exposure trends over time, and exposure profiles by occupation and agent are provided by FINJEM. Figure 5-1 shows examples of chemical exposure profiles (Kauppinen et al., 2014). To identify possible hazardous exposures by occupation and agent, these exposure estimates are compared with Finnish occupational exposure limits (Kauppinen et al., 2014).

FINJEM's trend information and exposure estimates have also been used to explore the extent of potential future attributable cases in Finland and have also been used in a European project that is examining the potential consequences of changing the occupational exposure limits of some carcinogens (Rushton et al., 2008; Kauppinen et al., 2014). The European project primarily used the methods for risk assessment developed in the United Kingdom and the industry-based CARcinogen EXposure (CAREX) system approach.

An analysis of the trends and potential trends of work-related exposure to 41 chemical agents in Finland from 1950 to 2020 used the following four metrics to calculate national exposure: the prevalence of exposure, the prevalence of high levels of exposure, the average level of exposure, and the national occupational inhalation exposure (Kauppinen et al., 2013). An estimate of dermal exposure was determined based on the number and extent of occupational skin diseases. From 1950-1970, it was found that there was a general pattern of increased exposure, followed by a decrease from 1970-2008 with an expected decrease to continue from 2008 and onward (Kauppinen et al., 2014).

The developers of FINJEM also pioneered efforts to account for psychosocial stressors in job exposure matrices. The FINJEM was used in 2004 successfully to examine psychosocial work stressors and dementia in Germany (Seidler et al., 2004).

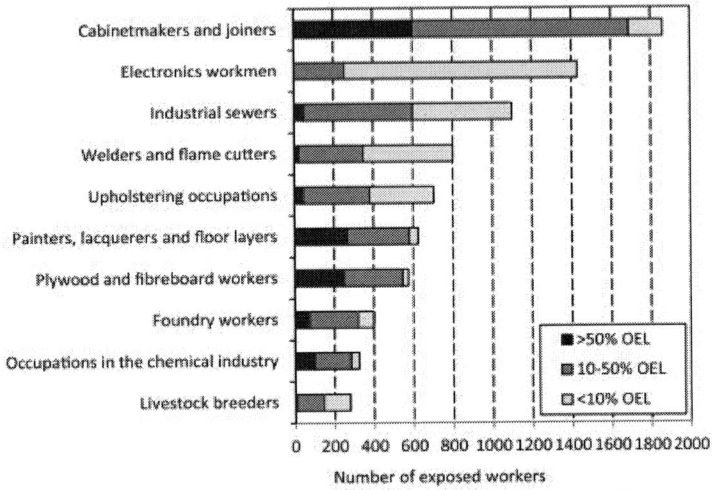


2 Occupational exposure profile for formaldehyde in Finland in 2007–2009. The numbers of workers exposed to formaldehyde by occupation and average level of exposure compared with Finnish OEL in 2009 (0.3 p.p.m.). Occupations with the 10 highest numbers of exposed workers are shown in the figure.

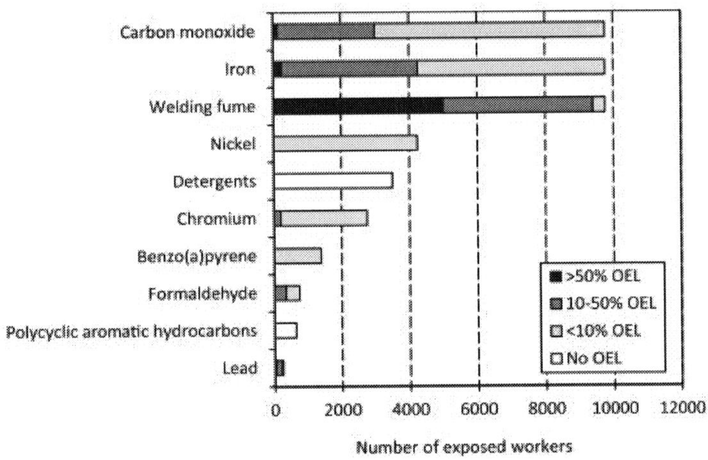


3 Occupational exposure profile for welders and flame cutters in Finland in 2007–2009. Numbers of workers exposed to chemical agent and average level of exposure compared with the Finnish OELs in 2009. Chemical agents with the 10 highest numbers of exposed workers are shown in the figure.

**FIGURE 5-1** Examples of occupational chemical exposure profiles provided in Finland. SOURCE: Kauppinen et al., 2014.

Job exposure matrices (JEMs) have also been used to assess other national trends in work-related exposures and hazards. In 2008, a Danish study used a JEM developed from the Danish Work Environment Cohort Study in a large case control study to show psychosocial job stressors were associated with anxiety disorders and with depression showing some variation by sex (Wieclaw et al, 2008). A record linkage study that linked a French JEM to a decennial health survey found a robust association of job strain with depressive symptoms (Cohidon et al, 2012). Most recently Australian investigators have validated a JEM for psychosocial stressors comparing assignments from the JEM with labor force survey responses for job stressors (Milner et al., 2016).

### European Working Conditions Survey

The European Working Conditions Survey (EWCS) conducted by Eurofound since 1991 (also discussed in Chapter 6) includes the exploration of a range of working conditions including physical and social environment. The EWCS is a household survey of populations of adults selected by a multistage, stratified, random sample of the working population in each country. For most countries, the sample size is 1,000 although larger samples come from several countries with large working populations. The sixth survey was conducted in 2015 and interviewed approximately 44,000 workers, which included both employees and the self-employed. This survey included 35 European countries, which consists of 28 EU member states, 5 EU candidate countries, and Norway and Switzerland (Eurofound, 2016). Survey questions included questions related to employment status, work organization, work-life balance, health and safety, learning and training, working time duration and organization, and physical and psychosocial risk factors (Eurofound, 2016). Seven job-quality indices, representing different dimensions of job quality, are assessed: physical environment, work intensity, working time quality, social environment, skills and discretion, prospects, and earnings. The questionnaire answers allow respondents to be classified into one of ten major occupational groups and one of ten major industry groups. Results show that in the past 10 years, there has been limited progress towards improvement in some job-quality indices (Eurofound, 2016).

### Reporting of Injuries, Diseases, and Dangerous Occurrences Regulations

As introduced above, the United Kingdom's RIDDOR program examines dangerous occurrences in the workplace that may require prompt reporting. These fall into three general categories:

- An event that occurs at any workplace "involving lifting equipment, pressure systems, overhead electric lines, electrical incidents causing explosion or fire, explosions, biological agents, radiation generators and radiography, breathing apparatus, diving operations, collapse of scaffolding, train collisions, wells, and pipelines or pipeline works" (HSE, 2017f).
- Incidents occurring at any onshore workplace that involve structural collapses, explosions or fires, releases of flammable liquids and gases, and hazardous escapes of substances.
- Incidents that have characteristics that are separately specified for reporting when they occur at offshore workplaces, at mines or quarries, or at relevant transport systems (HSE, 2017e).

## Conclusions

This examination of experience from other nations informs the efforts to develop a smart surveillance system for the United States. While the social systems and the political economies among the countries described differ in important ways, the lessons learned offer insights into how the United States might adapt aspects of the systems described to fill some of the gaps in our current approaches to surveillance of occupational injury and illnesses.

Fatal injury surveillance appears to be one area where the United States leads the way. The CFOI has evolved into the most comprehensive and specific system for identifying occupational fatalities with sufficient information to target intervention efforts. Surveillance of occupational commuting fatalities, however, is one area that has not been considered in the United States. Australia's experience suggests that surveillance of these events is possible and deserves consideration in the United States.

**Conclusion: Except for no access to CFOI data for case follow-up, this system for occupational fatality surveillance is internationally the most advanced.**

Non-fatal occupational injury surveillance, by contrast, is better developed in other countries. Employer reports elsewhere suffer similar problems to those in the U.S. Survey of Occupational Injuries and Illnesses, with underreporting a common problem. However, other nations rely more commonly on injury information available within national workers' compensation systems. For the most part workers' compensation systems in the United States are private and it would be difficult to adapt these to the same purpose. On the other hand, the success of the Safety and Health Assessment and Research for Prevention (SHARP) surveillance program in Washington State provides an excellent case example of how translating the international experience to the United States has been possible.

A common source of surveillance information on nonfatal injuries in other nations comes from labor force surveys. According to the HSE, this is the preferred source of information for occupational injuries in the UK. The Household Survey of Occupational Injuries and Illnesses (HSOII) proposed by the Bureau of Labor Statistics (BLS) is equivalent to a labor force survey although the survey is still in the planning stages with feasibility assessments under way.

**Conclusion: Occupational injury surveillance in the United States lags that of other developed nations.** This is in part due to the absence of a national workers' compensation system and in part due to the absence of a comprehensive labor force survey, both which are common in many developed nations.

Surveillance of occupational disease is challenging in all countries. This is recognized to be a consequence of the universal problem of poor training of physicians in the recognition of the role of occupation in the etiology of or as a contributing cause to disease. In many countries, the surveillance of occupational disease is based on a "schedule" of occupational diseases. For those patients found to have specific jobs or exposures associated with a specific list of diseases, the presumption is the condition is occupational in origin and is to be reported as such. The schedule concept serves to facilitate allocation of health care costs to the workers' compensation system. There is no comparable system in the United States although it might be possible to consider this approach on a state-by-state basis.

The UK model has adopted a somewhat different approach that offers some advantages and deserves scrutiny. The HSE requires direct reporting of eight specific conditions as well as reports of specific cancers or acute infectious disease when associated with known causes. This is supplemented by a voluntary physician reporting system (THOR) that has proved quite effective for select conditions. Among these are mental illnesses, conditions that are on the rise among the working population in the UK and likely in the United States. The Sentinel Event Notification System for Occupational Risks (SENSOR) system in the United States (see Chapter 4) has used variants on this system for asthma and pesticide poisoning and further developments in such an approach that could be given serious consideration.

**Conclusion: Occupational disease surveillance in most developed countries relies on a "schedule" of diseases that are presumed occupational if the schedule's known exposures are present.** There is no equivalent in the United States. The voluntary physician reporting scheme in the UK offers promise and the United States has limited experience in utilizing this type of surveillance.

Hazard and exposure surveillance, especially in Europe, has made important advances, and offers promising opportunities for the United States. One approach uses job exposure matrices to group jobs into exposure categories allowing an estimation of the distribution and location of priority hazards. Another includes the use of household surveys constructed to assess work and work exposures that can be effectively identified through worker self-reports. These have provided useful data on trends for common exposures among work populations for over two decades. Finally, there is the UK RIDDOR system that requires reporting of "dangerous occurrences without injury," certain specified near-miss events that are well characterized in HSE documents.

**Conclusion: Exposure surveillance efforts in other developed nations have proved useful for surveillance and offer significant promise for adoption in the United States.**

There are several approaches used by other countries that offer case studies of approaches to occupational injury and illness reporting that could be adapted to the U.S. setting. The experience is rich with examples that can serve to guide exploration of enhanced and more robust surveillance practices in the United States toward the further development of a smart surveillance system.

## REFERENCES

Agius, R., and S. Turner. 2004. Work-related mental ill health and development of the Surveillance of Occupational Stress and Mental Illness (SOSMI) reporting scheme. *Psychiatric Bulletin* 28:174-176.

Armstrong, H., and K. Bunn. 2012. Experience rating and occupational disease: A New Zealand case study. *Policy and Practice in Health and Safety* 10(1):63-75.

Australian Bureau of Statistics. 2013. Occupational injuries and disease. Chapter 15 in *Labour Statistics: Concepts, Sources and Methods, 2013*. Available online at http://www.abs.gov.au/ausstats/abs@.nsf/Lookup/6102.0.55.001Chapter252013 (accessed June 19, 2017).

AWCBC (Association of Workers' Compensation Boards of Canada). 2017. *About AWCBC*. Available online at http://awcbc.org/?page_id=10 (accessed June 19, 2017).

Benavides, F. G., G. L. Delcios, S. P. Cooper, and J. Benach. 2003. Comparison of fatal occupational injury surveillance systems between the European Union and the United States. *American Journal of Industrial Medicine* 44(4):385-391.

Carder, M., L. Hussey, A. Money, M. Gittins, R. McNamee, S. J. Stocks, D. Sen, and R. M. Agius. 2017. The Health and Occupation Research Network: An evolving surveillance system. *Safety and Health at Work*. 8(3):231-236.

Cohidon, C., G. Santin, J. F. Chastang, E. Imbernon, and I. Niedhammer. 2012. Psychosocial exposures at work and mental health: Potential utility of a job-exposure matrix. *Journal of Occupational and Environmental Medicine* 54(2):184-191.

Driscoll, T., A. Mannetje, E. Dryson, A.-M. Feyer, P. Gander, S. McCracken, N. Pearce, and M. Wagstaffe. 2004. *The burden of occupational disease and injury in New Zealand*. Available online at http://psm-dm.otago.ac.nz/ipru/ReportsPDFs/OR057.pdf (accessed May 5, 2017).

Eurofound. 2016. *Sixth European Working Conditions Survey: Overview report*. Available online at https://www.eurofound.europa.eu/sites/default/files/ef_publication/field_ef_document/ef1634en1_0.pdf (accessed June 22, 2017).

EUROGIP. 2015. *Reporting of Occupational Diseases: Issues and Good Practices in Five European Countries*. Paris: EUROGIP. Available online at http://www.eurogip.fr/images/documents/3933/Report_DeclarationMP_EUROGIP_102EN.pdf (accessed June 21, 2017).

European Agency for Safety and Health at Work. 2017. *Commission recommendation concerning the European schedule of occupational diseases*. Available online at https://osha.europa.eu/en/legislation/guidelines/commission-recommendation-concerning-the-european-schedule-of-occupational-diseases (accessed June 19, 2017).

European Commission. 2013. *Report on the current situation in relation to occupational diseases' systems in EU member states and EFTA/EEA countries, in particular relative to commission recommendation 2003/670/EC concerning the European schedule of occupational diseases and gathering of data on relevant related aspects*. Available online at https://osha.europa.eu/en/legislation/guidelines/commission-recommendation-concerning-the-european-schedule-of-occupational-diseases (accessed June 21, 2017).

Eurostat. 2017a. *Eurostat methodology*. Available online at http://ec.europa.eu/eurostat/web/health/methodology (accessed May 8, 2017).

Eurostat. 2017b. *European Union labour force survey*. Available online at http://ec.europa.eu/eurostat/web/microdata/european-union-labour-force-survey (accessed June 19, 2017).

Guidotti, T. L. 2013. Occupational diseases. In *Canadian Encyclopedia*. Available online at http://www.thecanadianencyclopedia.ca/en/article/occupational-diseases (accessed June 21, 2017).

Hours, M., E. Fort, B. Charbotel, and M. Chiron. 2011. Jobs at risk of work-related road crashes: An analysis of the casualties from the Rhone Road Trauma Registry (France). *Safety Science* 49(8-9):1270-1276.

HSE (Health and Safety Executive). 2016a. *Work-related skin disease in Great Britain, 2016*. Available online at http://www.hse.gov.uk/statistics/causdis/dermatitis/skin.pdf (accessed June 21, 2017).

HSE. 2016b. *Work-related respiratory disease in Great Britain, 2016*. Available online at http://www.hse.gov.uk/statistics/causdis/respiratory-diseases.pdf?pdf=respiratory-diseases (accessed June 21, 2017).

HSE. 2017a. *About the Labour Force Survey (LFS)*. Available online at http://www.hse.gov.uk/statistics/lfs/about.htm (accessed June 19, 2017).

HSE. 2017b. *Table of preferred sources for injuries and ill health*. Available online at http://www.hse.gov.uk/statistics/preferred-data-sources.htm (accessed June 19, 2017).

HSE. 2017c. *Data sources*. Available online at http://www.hse.gov.uk/statistics/sources.htm (accessed May 8, 2017).

HSE. 2017d. *Reportable diseases*. Available online at http://www.hse.gov.uk/riddor/occupational-diseases.htm (accessed June 19, 2017).

HSE. 2017e. *RIDDOR—Reporting of Injuries, Diseases and Dangerous Occurrences Regulations 2013*. Available online at http://www.hse.gov.uk/riddor (accessed June 22, 2017).

HSE. 2017f. *Dangerous occurrences*. Available online at http://www.hse.gov.uk/riddor/dangerous-occurences.htm (accessed December 4, 2017).

Hussey, L. 2016. *THOR-GP input to the HSE annual statistics, 2015/2016*. Available online at http://www.hse.gov.uk/statistics/calculation-thor-gp-data.pdf (accessed June 19, 2017).

Hussey, L., M. Carder, A. Money, and R. Agius. 2013. Comparison of work-related ill-health data from different GB sources. *Occupational Medicine* 63(1): 30-37.

ILO (International Labour Organization). 2010. *ILO list of occupational diseases (revised 2010)*. Available online at http://www.ilo.org/safework/info/publications/WCMS_125137/lang--en/index.htm (accessed June 19, 2017).

Japan Ministry of Health, Labour, and Welfare. 2017. *Survey on industrial accidents*. Available online at http://www.mhlw.go.jp/english/database/db-l/industrial_accidents.html (accessed May 8, 2017).

Jarosz, B., and R. T. Cortes. 2014. *In U.S., new data show longer, more sedentary commutes. Population Reference Bureau article*. Available online at http://www.prb.org/Publications/Articles/2014/us-commuting.aspx (accessed June 19, 2017).

JISHA (Japan Industrial Safety & Health Association). 2017. *Industrial accidents statistics in Japan, 2010. Statistics in Japan*. Available online at http://www.jisha.or.jp/english/statistics/accidents2010.html (accessed May 8, 2017).

Kauppinen, T., S. Uuksulainen, A. Saalo, and I. Mäkinen. 2013. Trends of occupational exposure to chemical agents in Finland, 1950-2020. *Annals of Occupational Hygiene* 57(5):593-609.

Kauppinen, T., S. Uuksulainen, A. Saalo, I. Mäkinen, and E. Pukkala. 2014. Use of the Finnish Information System on Occupational Exposure (FINJEM) in epidemiologic, surveillance, and other applications. *Annals of Occupational Hygiene* 58(3):380-396.

Kneebone, E., and N. Holmes. 2015. *The growing distance between people and jobs in metropolitan America*. Available online at https://www.brookings.edu/research/the-growing-distance-between-people-and-jobs-in-metropolitan-america (accessed June 19, 2017).

Milner, A., I. Niedhammer, J. F. Chastang, M. J. Spittal, and A. D. LaMontagne. 2016. Validity of a job-exposure matrix for psychosocial job stressors: Results from the Household Income and Labour Dynamics in Australia Survey. *PLoS ONE* 11(4): e0152980.

Munich Re Group. 2004. *Commuting accidents: A challenge for workers' compensation systems*. Available online at http://s3.amazonaws.com/zanran_storage/www.munichre.com/ContentPages/19215127.pdf (accessed June 19, 2017).

Pearce, N., E. Dryson, A. M. Feyer, P. Gander, S. McCracken, and M. Wagstaffe. 2005. *Surveillance of Occupational Disease and Injury in New Zealand: Report to the Minister of Labour*. Wellington, New Zealand: Occupational Health and Safety Advisory Committee. Available online at http://psm-dm.otago.ac.nz/ipru/ReportsPDFs/OR056.pdf (accessed May 5, 2017).

Rushton, L., S. Hutchings, and T. Brown. 2008. The burden of cancer at work: Estimation as the first step to prevention. *Occupational and Environmental Medicine* 65(12):789-800.

Safe Work Australia. 2017a. *Work-related fatalities*. Available online at https://www.safeworkaustralia.gov.au/statistics-and-research/statistics/fatalities/fatality-statistics (accessed June 19, 2017).

Safe Work Australia. 2017b. *Explanatory notes for Safe Work Australia datasets*. Available online at https://www.safeworkaustralia.gov.au/system/files/documents/1702/explanatory-notes-datasets.pdf (accessed June 19, 2017).

Seidler, A., A. Nienhaus, T. Bernhardt, T. Kauppinen, A. L. Elo, and L. Frolich. 2004. Psychosocial work factors and dementia. *Occupational and Environmental Medicine* 61:962-971.

Statistics New Zealand. 2014. *Serious injury outcome indicators: 2000–13*. Available online at http://www.stats.govt.nz/browse_for_stats/health/injuries/serious-injury-outcome-indicators-2000-13.aspx (accessed June 15, 2017).

Stocks, S. J., S. Turner, R. McNamee, M. Carder, L. Hussey, and R. M. Agius. 2011. Occupation and work-related ill-health in UK construction workers. *Occupational Medicine (London)* 61(6):407-415.

Takahashi, K., and Y. Ishii. 2014. *Historical Developments of Administrative Measures for Occupational Diseases in Japan*. Geneva: International Labor Organization. Available online at http://www.ilo.org/safework/info/publications/WCMS_234221/lang--en/index.htm (accessed June 15, 2017).

Tedone, T. S. 2017. Counting injuries and illnesses in the workplace: An international review. *Monthly Labor Review*, September. Available online at https://www.bls.gov/opub/mlr/2017/article/counting-injuries-and-illnesses-in-the-workplace.htm (accessed December 21, 2017).

Walters, D. 2007. *An International Comparison of Occupational Disease and Injury Compensation Schemes*. Report. Cardiff Work Environment Research Centre, Cardiff University. Available online at https://www.gov.uk/government/uploads/system/uploads/attachment_data/file/330347/InternationalComparisonsReport.pdf (accessed September 20, 2016).

Warburton, K. L., R. Urwin, M. Carder, S. Turner, R. Agius, and S. M. Wilkinson. 2015. UK rates of occupational skin disease attributed to rubber accelerators, 1996-2012. *Contact Dermatitis* 72(5):305-311.

Wieclaw, E., E. Agergo, P. B. Mortensen, H. Burr, F. Tuchsen, and J. P. Bonde. 2008. Psychosocial working conditions and the risk of depression and anxiety disorders in the Danish workforce. *BMC Public Health* 8:280.

WorkSafe New Zealand. 2017. *Sources of work-related injury data*. Available online at http://www.worksafe.govt.nz/worksafe/research/health-and-safety-data/sources-of-work-related-injury-data (accessed May 8, 2017).

Zhou, A. Y., M. Carder, L. Hussey, M. Gittins, and R. Agius. 2017. Differential reporting of work-related mental ill-health in doctors. *Occupational Medicine* 67(7):522-527.

# 6

# Promising Developments and Technologies

## INTRODUCTION

The preceding chapters have described a set of objectives and a shared vision for a "smart" occupational safety and health (OSH) surveillance system and have provided an overview of the current status of occupational safety and health surveillance as well as recommendations to strengthen those efforts. This chapter discusses ongoing efforts that offer promise for improving occupational health and safety surveillance. Seven areas are focused on:

- Exploration and implementation of a household survey,
- Use of electronic health records,
- Coding of occupational information,
- Electronic reporting initiatives,
- Use of workers' compensation data,
- Leveraging existing surveys and data systems, and
- Improving occupational hazard and exposure surveillance.

Each topic is explored briefly and that description is followed by the committee's recommendation on that issue.

This chapter focuses on a variety of activities that are already under way and that constitute opportunities for improving OSH surveillance. In the next chapter, the committee introduces the enabling components and emerging methods that may further enhance our ability to achieve a smarter system for OSH surveillance in the future.

## IMPLEMENTING A HOUSEHOLD SURVEY

The Bureau of Labor Statistics (BLS) is exploring the feasibility of conducting a nationwide Household Survey of Occupational Injuries and Illnesses (HSOII) with the goals of better understanding the needs of workers, employers, and the safety and health community and addressing the current undercount of occupational illnesses and injuries (see Chapter 4). The HSOII would contact workers directly—outside of the employer-employee relationship—and would ask questions similar to the establishment-based (i.e., employer-based) Survey of Occupational Injuries and Illnesses (SOII) to provide additional data and allow for comparability (Monaco, 2016). The primary goal of the new survey would be to produce as complete as possible a measure (counts and rates) of occupational injuries and acute illnesses in the U.S. economy by capturing *all* workers including self-employed, contract and "gig" workers (defined in Chapter 1), household workers, migrant laborers, and immigrant workers. Additionally, the new survey would identify gaps in the estimates derived from the SOII. The household survey under development has been designed on a statistically valid platform that will provide better demographic data as well as present the opportunity to ask occupational safety and health questions directly of employees, facilitate special studies, allow for rotating topics and questions, and obtain improved descriptions of acute events. Nonresponse will likely be a challenge, thus BLS has proposed several methods to improve response rates for the household survey, including the use of dialing protocols that rotate calls through different times of day and days of week, increasing the number of attempts of contact, and maximizing use of highly-trained interviewers (https://www.reginfo.gov/public/do/DownloadDocument?objectID=71013400).

This approach would build on BLS's successful record of collaborating on the collection of complicated and sensitive information using population surveys, such as its collaboration with the Bureau of Census on the Current Population Survey (CPS). The basic infrastructure for such a survey is already in place and would provide results that would be consistent over time and across states, as well as being statistically valid and providing a measure of reliability. As currently envisioned in a recent report produced for BLS, the HSOII would not be sufficiently robust to allow examination of findings that are state specific. The target population being considered for an HSOII is workers age 16 years and older, with a worker being anyone who worked in at least 1 of the prior 52 weeks (NORC, 2016b).

BLS contracted with the National Opinion Research Center (NORC) to develop recommendations for survey design options intended to meet the requirements of such a survey regarding sample representativeness, data quality, timeliness, and cost (NORC, 2016a,b). NORC provided three options for a HSOII (NORC, 2016b). The one that is considered most economical—and assessed by NORC as meeting the full requirements—is the option that uses "supplemental questions on occupational safety and health following the CPS Annual Social and Economic Survey (sometimes referred to as the CPS March Supplement) for those sample persons identified as meeting the HSOII eligibility requirements" (NORC, 2016b). About 3 percent of CPS respondents would be asked the full HSOII questionnaire under this option (NORC, 2016b).[1] A second option would add questions to the June or July CPS when there is no major supplement, but this option is slightly costlier than the first as it entails slightly more screening and respondent burden. A third option would use the ACS respondents as a sampling frame, and then administer the full HSOII separately from the ACS. This option offers more flexibility in targeting selected industries and occupations, not possible with the first two options that add questions to the CPS, but at a higher cost, due to increased respondent burden and additional fixed costs of administering a separate survey rather than adding to an existing survey.

According to NORC's estimates, the first two options based on adding questions to the CPS would produce 51,000 to 57,000 completed interviews within a $1 million budget, while the third option would yield less than 50,000 completed interviews within the same budget (NORC, 2016b). The third option, therefore, represents a trade-off between greater flexibility in targeting the HSOII versus a higher cost per completed interview. Before selecting one of these options, BLS plans to conduct a pilot survey on a smaller, nationally representative sample of about 5,000 workers to assess their occupational injuries and illnesses, with the goal to have results available in 2018 (Monaco, 2016). In this pilot, the screening questions in the core CPS will serve to identify the subset of respondents to receive the full HSOII questionnaire—those aged 16+, working, and who reported in the core CPS that they had sustained a work-related illness or injury.

A similar effort to assess workplace injuries by self-report has been carried out in the United Kingdom: the Labour Force Survey (see Box 6-1). The United Kingdom's Health and Safety Executive (HSE) uses two sources of data to assess occupational injury and illness: (1) the Reporting of Injuries, Diseases and Dangerous Occurrences Regulations (RIDDOR), the equivalent of the SOII, and (2) the Labour Force Survey, akin to the proposed

[1]In addition to the ~57,000 CPS ACES households, of which ~65% of the individuals in the sample aged 16+ are estimated to be working, consideration is being given to include the ASEC supplemental samples of 6,500 Hispanic HUs and 19,000 CHIP HUs (NORC, 2016b). Based upon CPS response rates and ACS worker rates, it is estimated that this inclusion would add an additional 23,400 sample persons eligible for the HSOII (NORC, 2016b). Consequently, the sample yield estimated to receive the full set of HSOII questions should substantially exceed 4,000 (NORC, 2016b).

**BOX 6-1**
**Overview of the Labour Force Survey**

The Labour Force Survey in the United Kingdom is based on a household survey covering approximately 38,000 households each quarter. Among the questions asked in this survey are whether an injury was a result of a work-related accident[1] during the past 12 months, whether injury was caused during driving, and how soon after the event did the individual return to work. From these data, the HSE computes injury rates for all work-related accidents and injury rates from "reportable" workplace accidents (>3 day work absence and >7 day work absence; the latter is the same as what is required to be reported to the UK RIDDOR–equivalent of SOII).

The Labour Force Survey is seen to address the underreporting from employer reports, as well as excluded workers such as the self-employed. As currently designed the survey is recognized to have limitations, including absence of data on nature and severity of the injury (although the belief is that self-reports are dominated by <3 day work absence events), recall bias (as the question asks for any injury in the past 12 months), and difficulty in referencing injuries to current jobs for those who have changed jobs in the past 12 months.

This survey is also regarded as an important source of information about work-related illness. Respondents to this survey are asked to report conditions they believe have been caused or made worse by work. Useful analysis of these data provides both prevalence and incidence information about occupationally related conditions. These data have been collected periodically since 1990 and annually from 2001-2002. The most common reports of illness from this system include stress, musculoskeletal disorders, asbestos-related disease, respiratory disease, deafness, skin disease, and vibration-related disease.

[1] In U.S. surveys of work injuries, the question is asked about work-related injuries or injuries related to a job. The term "accident" is no longer used.
SOURCE: HSE, 2017.

HSOII. These two systems are considered complementary. The HSE does not attempt to integrate the two directly to arrive at a single annual estimate of occupational injury and illness. Rather they present both rates along with analysis of determinants of the rates based on data available from each of the systems.

As with the United Kingdom's two surveys, the working design for the HSOII targets acute injuries and illnesses in a manner meant to parallel the objectives of the SOII. Despite being significantly challenged in terms of enhancing the details of injuries as currently represented in the SOII, the HSOII will serve a critically important function by improving the count and hence the understanding of the burden of injuries in populations not covered by the current SOII. The secondary goal—using this approach to develop a better understanding of the undercount—may be a greater challenge to fulfill. Recordkeeping requirements from the Occupational Safety and Health Administration (OSHA) provide the basis for the SOII and include parameters such as days lost, treatment beyond first aid, and others. BLS has determined that these parameters are currently not well understood by employers, therefore, there are likely to be significant challenges to formulating survey questions on these aspects of an injury in such a way that they are understood by individual survey respondents.

Two additional gaps recognized in the 1987 National Research Council (NRC) report could potentially be addressed in the household survey:

- The almost complete absence of information about chronic diseases that are either uniquely caused by work or that include work as one of several important factors, and
- The need to develop and enhance information about hazards at work.

Opportunities for collecting self-reported information on chronic conditions was discussed in Chapter 4. Although the quality of information on disease that is self-reported is always potentially problematic, the National Center for Health Statistics (NCHS) has been successful in collecting and reporting on reliable information from respondents asked about their health. For example, in the ongoing National Health Interview Survey (NHIS), conditions such as low back pain, carpal tunnel syndrome, eye-nose-throat irritation, and skin conditions have been reliably reported by respondents, as has information about doctor-diagnosed diseases such as diabetes and high blood pressure. With insights from the NCHS experience regarding disease and chronic health conditions, BLS could consider including these questions in the HSOII.

BLS would also need to consider the HSOII's potential to collect information about work environment hazards that have not been available nationally since the National Institute for Occupational Safety and Health (NIOSH) stopped conducting the National Occupational Hazard/Exposure Surveys in the early 1980s (see Chapter 4 and the section on "approaches to hazard and exposure surveillance" later in this chapter). The HSOII presents an opportunity to collect self-reports of important and prevalent working conditions, including exposure to physical hazards (exposure to noise, dust, chemicals, or infectious agents; lifting heavy loads; and repetitive hand movements), work intensity (working at speed and to tight deadlines, not having enough time to do the job, frequent disruptive interruptions, pace determinants and interdependency, and emotional demands), types of working relationships (shift work, working hours, second jobs, working alone, and contract or on-demand work), and skills and discretion (cognitive dimension of work, decision latitude, organizational participation, access to training, use of technology at work, and teamwork). Such information has been collected and tracked successfully in the European Working Conditions Survey approximately every 5 years since 1991[2] (Eurofound, 2017a). Work organization factors currently have been collected in NHIS supplements on occupational health in 2010 and 2015 (NCHS, 2016), and the importance of collecting exposure information on psychosocial factors has been provided by Australian researchers concerned with important factors at work that affect mental health (LaMontagne et al., 2016). Enhancement of HSOII by questions about health conditions and disease and work exposures would require further careful investigation concerning respondent burden and feasibility.

The results of the household survey (HSOII) and employer-based survey (SOII) would need to be disseminated together as they would offer complementary insights. A nationwide HSOII expanded to include health conditions and diseases as well as exposure to work hazards would need to occur at least every 5 years or more frequently if feasible. Some health conditions that are well known to be caused by work (e.g., silicosis and lung cancer) have long latency periods and so do not appear in SOII. Many other health conditions are not uniquely caused by work but work is one important factor in the development and evolution of disease. While it would be desirable to have these conditions reported and tracked annually, a survey interval of 5 years would add immeasurably to the knowledge base of the distribution and determinants of these conditions.

**Conclusion: A household survey on occupational injuries and illnesses would provide data needed to provide more comprehensive surveillance data that will include important information currently lacking in the SOII.** The committee finds that the HSOII will serve the BLS objectives to provide

- Greater accuracy by capturing data on all workers in the U.S. economy;
- A statistically valid platform to ask occupational safety and health questions through special studies, rotating topics and questions, better demographic data, and improved description of the event; and
- Self-reported data on occupational injuries or illnesses that provide

[2]The European Foundation for the Improvement of Living and Working Conditions's (Eurofound's) 2015 European Working Conditions Survey was administered in 35 countries to nearly 44,000 subjects through computer-assisted interviews conducted in 49 languages (Eurofound 2017a,b).

a complement to the employer-based survey (SOII) so that together the two sources offer broader insights to prevention.

In addition to these advantages, experience from the United Kingdom and from the European Foundation for the Improvement of Living and Working Conditions [Eurofound] provide strong evidence that the HSOII approach presents an excellent opportunity to collect information that has been missing from any routine national surveys:

- Reports of chronic conditions or diseases and their relationship to work, and
- Reports of a wide variety of work hazards that cover many conditions that are considered important in affecting the acute and chronic health of the workforce.

**Recommendation D: BLS should place priority on implementing its plan for a household survey of nonfatal occupational injury and illnesses (HSOII).** With the assistance of NIOSH and the Centers for Disease Control and Prevention (CDC), BLS should also expand this effort to include a periodic nationwide household survey to identify and track reports of occupational exposures and should determine how best to identify and track chronic work-related illnesses.

In the near term:

- BLS should survey occupational injuries and acute illnesses (as in SOII) in a nationally representative sample of the entire working population including those who are self-employed or engaged in temporary contract work.

In the longer term:

- To address the inadequacies of current surveillance tools, BLS should
  - Seek assistance from NIOSH to enhance the HSOII survey scope by assessing occupational exposures and risks in a manner like that used in the Eurofound Working Conditions Survey.
  - Questions should be included to capture exposure determinants and work characteristics with sufficient details on industry, occupation, work organizational characteristics, and working relationships in a way that supports the development of a flexible job exposure matrix and supports integration of newly available or ancillary data.
  - Seek assistance from NCHS and NIOSH to address currently

inadequate information on chronic disease and work by determining whether self-report of illnesses and chronic conditions is best tracked by inserting occupational information into the NHIS or inserting chronic illness questions into the HSOII. Part of this consideration should include the determination of whether a sample of retirees and those not working due to disability should be part of the HSOII.

- BLS should prepare and implement a specific plan for routine analysis, interpretation, and preparation of a report on the findings from the HSOII along with a plan for dissemination and appropriate database access by researchers and the public.

## ELECTRONIC HEALTH RECORDS

Passage of the 2009 Health Information Technology for Economic and Clinical Health (HITECH) Act has led to increased adoption of electronic health records (EHRs) by both individual clinicians and health care provider organizations. With the goal of ensuring the *meaningful use* of the EHR, the Office of the National Coordinator for Health Information Technology (ONC) is guiding a process that provides financial incentives so that EHR designers and vendors will develop new EHRs, and augment older ones, to meet a broad range of requirements that enhance the role of information technology in supporting clinical and population health (HealthIT, 2017). The meaningful use criteria established by the ONC are designed to result in better clinical outcomes, improved population health outcomes including reduction in health disparities, increased transparency and efficiency, empowered individuals, and more robust and integrated research data available through health data systems.

The increasing adoption of the EHR and the establishment of meaningful use criteria offer an unprecedented opportunity to improve data capture of the impact of work and work exposures on individual health. Advances in capturing these data could improve occupational health practice and are expected to inform general medical practice by more effectively placing health conditions in the context of work. This context is particularly important when accounting for the impact of work on a health condition, even if the condition was not caused by work (for example, a diabetic working rotating shifts needs to plan to adjust insulin management to the pattern of working hours). Readily accessible occupational data in health records also can inform community-health needs assessments. Every 3 years, under the Patient Protection and Affordable Care Act (P.L. 111-148), tax-exempt 501(c)(3) hospitals are required to conduct a Community Health Needs Assessment to develop and adopt an implementation plan to address an identified population health need. Local health departments seeking national accreditation must also conduct a Community Health Needs Assessment every 5 years as part of their strategic planning.

NIOSH has been engaged in efforts to promote the inclusion of occupational information for health in EHRs (MacKenzie et al., 2016). Inclusion of occupational data in the EHR offers promise for improved clinical care but also would enhance surveillance of occupational disease and injury to improve population health and address health disparities through several routes. First, data from the EHR could be used to improve documentation of injuries related to work. Second, the data in the EHR could potentially be used to identify when occupation is partially or wholly responsible for illness or disease (i.e., asthma in a machinist or chronic obstructive pulmonary disease in a miner), especially chronic conditions. And third, the EHR data could be used to identify work or work-exposure relationships with new or emerging patterns of disease. Inclusion of occupational information in the EHR is critically important to increasing the availability of this information in other public health data systems used for surveillance such as cancer registries, trauma registries, and emergency department data sets such as the National Electronic Injury Surveillance Systems—Occupational Supplement (NEISS Work) that rely on obtaining this information from medical records.

Shortly after the enactment of the HITECH Act, NIOSH began an effort to establish meaningful use requirements for the collection of occupational data through an EHR. As part of this effort, NIOSH asked the National Academies to analyze the potential benefits of including occupational data in the EHR and how technical challenges for the effective incorporation of occupational data in the EHR could be overcome. The ensuing report of the Institute of Medicine (IOM, 2011) noted that, from the perspective of public health or occupational health surveillance, linking occupation data to patient care data provides an opportunity to evaluate injuries, illnesses, and health status in relation to work in the populations receiving care. The report recommended NIOSH focus initially on developing feasible means to incorporate an appropriate level of data on occupation, industry, and work-relatedness into the EHR and subsequently consider what efforts were needed to enhance the value and use of occupational data that would be available in the EHR in the future.

In response to the recommendations in that report, NIOSH undertook a series of projects to build support for the capture of occupational data using EHR systems primarily focused on demonstrating the feasibility of capturing these data in the record, modifying or developing guidance regarding software systems for efficient management and retrieval of occupational data in the electronic record, and developing model systems of clinical decision support for the specific examples of asthma, musculoskeletal disorders, and diabetes (Allen et al., 2017; Baron et al., 2017; Filios et al., 2017; Harber et al., 2107; McLellan et al., 2017).

The occupational data that minimally meet the needs identified in the IOM report are *current occupation and industry* (useful for acute occupational injury, short-latency occupational illnesses, and management of current medical conditions) and *usual occupation and industry* (necessary for evaluating occupational illnesses of long latency) (IOM, 2011).

Once in the EHR, current occupational information and, more importantly, the development over time of full occupational histories will permit examination of specific common conditions to seek important signals for an occupational factor. Once noted, these can be tracked, hot spots can be identified, and formal etiologic studies can be planned to advance knowledge and prevention.

A related concern is determining the optimal time and method for collecting occupational data during a clinical encounter. Currently collection generally occurs at the intake interview, if at all. Ensuring that occupational data are accurately entered in the EHR will require tools and training. Pilot efforts suggest that personnel can be effectively encouraged to collect these data for most jobs (NIOSH, 2017). As an alternative, NIOSH is assessing the feasibility of having patients enter occupational data themselves at the time of intake. This approach will most likely require a dictionary of job titles or industries in a drop-down list to facilitate selecting a standard term from a pick list, although it may also be possible to assign codes automatically to narratives entered by patients.

Regardless, for occupational data to be most useful, they must be linked to the clinical record when they are collected. NIOSH has thus focused its attention on interoperability, or the ability to move structured data between systems in a manner that uses standards that ensure that the data can be used by the recipient system. Such interoperability is required for effective data movement among patient care domains as well as between clinical domains and public health systems. Such efforts also pertain to systems that automatically encode data captured as free text and can utilize methods that provide common language translations of those codes for use by other systems (see section on "coding of occupational data" in this chapter).

NIOSH is also exploring what additional data would need to be collected about work. It has framed an information model, which could further support clinical care, population health, and public health activities by including information about work hours, schedule, job duties, and exposures (NIOSH, 2017).

Finally, for data collected in an EHR to be useful for public health surveillance, they must be accessible by the pertinent agencies. One possible mechanism is electronic case reporting, wherein logic is encoded in the EHR to identify cases of reportable conditions and then to report the cases automatically to the appropriate public health agency. Efforts are well under way in the United States to define a technical framework and data elements, including occupation, for electronic case reporting (MacKenzie et al., 2016).

### Summary and Conclusion

The 2011 IOM report provides extensive documentation about the value of routinely capturing occupational data in an EHR (IOM, 2011). Those findings are reinforced by work done since that report. Currently, the major impediments to progress are feasibility and the availability of resources.

In 2015, the ONC acknowledged that occupational data captured in the EHR can benefit patient care and population health, but noted that data standards and software tools for capturing these data in the EHR were too immature at that time to establish a certification criterion for the capture of occupational data (HHS, 2015a). The ONC has indicated that it will monitor the development of these tools and standards for future rulemaking (HHS, 2015b).

**Conclusion: Routine inclusion of occupational data in the EHR is required for improved diagnosis and treatment of work-related conditions; for better-informed management of health conditions that are affected by work circumstances; for enhanced understanding of community health needs and resources; and for local, state, and national surveillance of occupationally related conditions.** Further efforts are critical to ensure that industry and occupation are included in the EHR meaningful use data.

## CODING OF OCCUPATIONAL DATA

Data relevant to occupational health and safety surveillance have three basic dimensions: occupation, industry, and a description of the case characteristics of an occupational injury, illness, or fatality.

For occupation and industry, the official coding schemes or controlled terminology used by federal statistical agencies are the Standard Occupational Classification (SOC) for occupation and the North American Industry Classification System (NAICS) for industry. The U.S. Census Bureau has a related less detailed industry and occupation coding system, derived from SOC and NAICS, for purposes of coding data provided by individuals. The Census occupation codes and Census industry codes are mapped to the SOC and NAICS codes (see Box 6-2).

Workers' compensation insurance rating firms and agencies use classification systems that combine elements of occupation and industry. These

**BOX 6-2**
**Coding Systems for Occupation and Industry**

The Standard Occupational Classification (SOC) system is used by most federal statistical agencies to categorize workers into occupational categories. All workers are classified into one of 840 detailed occupations that are, in turn, grouped into 461 broad occupations, 97 minor groups, and 23 major groups (BLS, 2017a). The system is updated periodically with the next version expected in 2018.

The North American Industry Classification System (NAICS) is the standard used by federal statistical agencies to classify business establishments based on the type of business activity (process or production) for monitoring the economy. The NAICS divides the economy into 709 industries, which are, in turn, grouped into 311 industry groups, 99 subsectors, and 20 sectors (U.S. Census Bureau, 2017a). The classification system is updated every 5 years, most recently in 2017, with crosswalks linking the current classification to historical versions.

The U.S. Census Bureau codes occupation and industry using a slightly different set of codes for the American Community Survey and Current Population Survey, among others. These code lists are directly derived from SOC and NAICS but have fewer terms, reflecting the extent of what can be reliably collected in their surveys. These codes are mapped to the codes used in the SOC and NAICS systems (U.S. Census Bureau, 2017b).

The National Council on Compensation Insurance (NCCI) "manual classification" codes classify businesses for workers' compensation insurance purposes in states subscribing to NCCI services (NCCI, 2017). Workers' compensation codes vary across the states.

classifications are called "manual classes," with the most commonly used systems across the United States being those from the National Council of Compensation Insurance (an example from New York can be found at New York Workers' Compensation Board (NY WCB, 2017); also see Box 6-2 and discussion under workers' compensation section below).

Coding of type of work and workplace is complemented by classification systems to characterize the occupational injury or illness. Injury and illness classification is an important concern—the classification will likely vary based on who provides the information (health care provider, employer, or worker), what data are available (narrative text with injury descriptions, and physician diagnoses), when it is provided during the course of care, and the quality of the coders.

Injury or illness events are coded in three different ways depending on the source of reporting. There is no clear mapping between the terms or codes in the three systems (see Box 6-3).

Occupation and industry are often recorded as free text in death certificates, some birth certificates, and in several national surveys (e.g., NHIS, the Behavioral Risk Factor Surveillance System [BRFSS], the National Health and Nutrition Examination Survey [NHANES]), cancer registry reports, and clinical records). Currently, there is no single or universal standard rule or system that institutes a uniform approach to recording occupational data in these documents. For example, the U.S. standard certificate of death includes a section where the funeral director is asked to record the "decedent's usual occupation" ("indicate type of work done during most of working life; do not use retired") and "kind of business/industry." NIOSH

**BOX 6-3**
**Coding Systems for Injury and Illness Events**

The Occupational Injury and Illness Classification System includes four hierarchical coding structures: nature of the injury or illness, part of body affected, source and secondary source, and event or exposure. This system is used by BLS to code injuries or illnesses recorded in the SOII and the Census of Fatal Occupational Injuries (BLS, 2014a; CDC, 2017a).

The Workers Compensation Insurance Organizations have developed codes for nature of injury, cause of loss, and part of body affected when reporting workers' compensation injuries. Various workers' compensation entities include these codes as part of their reporting standard although this is not uniformly the case across all systems. Some of the codes may not be used by individual jurisdictions, or there may be variations in definitions for certain codes. The organization's Electronic Data Interchange Committee administers changes to the code on the reference list (WCIO, 2017).

The *International Classification of Diseases, 10th Revision* (ICD-10) is an international standard for classifying and presenting cause of death in mortality statistics. The ICD-10 Clinical Modification (CM) is used by health care providers for classifying and coding diagnoses, symptoms, and procedures recorded in conjunction with hospital care. The ICD-10 includes external cause of injury codes that provide detail on the environmental events, circumstances, and conditions as a cause of injury. An expanded set of external cause of injury codes are included in the ICD-10-CM, which includes a code for identifying injury during the course of paid civilian employment. There is no national requirement for mandatory external cause recording although some states require such coding and providers can voluntarily report external causes. The ICD-11 is expected to be released in 2018 (WHO, 2016).

has developed guidance for funeral home directors concerning completion of this section (NIOSH, 2012). The effectiveness of this guidance in improving quality and detail of occupational information across the 50 states has not been evaluated. However, the form of the question about work and the instructions for completing this section both vary among the 50 states as not all states follow the U.S. standard. Regardless, the information on occupation and industry in death certificates is present more than 95 percent of the time and has proved useful for surveillance and research (NIOSH, 2012).

Use of these data for surveillance requires that data written as free text be translated into codes appropriate for the various record types. The process of extracting structured data (or codes) from unstructured data (or free text) requires the following:

- The existence of one or more defined coding system(s),
- Trained individuals and/or software to extract the codes from the free text, and possibly
- The incorporation of the coding system(s) and software into a larger system (e.g., an EHR).

Several coding software systems have been designed to extract a common set of codes for occupation and industry from free text. NIOSH has developed and continues to enhance the NIOSH Industry and Occupation Computerized Coding System (NIOCCS), a coding software package designed to extract injury and occupation codes from free-text data (NIOSH, 2016a). This web-based system is designed to map free text to Bureau of Census codes for industry and occupation with a crosswalk function mapping those codes to NAICS and SOC codes, the major coding systems for most record types. Developed initially to code occupation and industry information on death certificates, NIOCCS has been expanded to other data sources. Thus far, NIOSH has been able to demonstrate moderate success with approximately 60 to 70 percent accuracy in coding industry and occupation by NIOCSS compared with coding by trained coders using a variety of different record types (NIOSH, 2016b). NIOCCS v.2 is currently available for public use and allows for single record or batch processing and for automatic as well as computer assisted coding. Continuous improvement has been documented and the NIOCSS software version 3.0 was scheduled to be released as of the release of this report (January 2018). At least one study that used NIOCCS to code cancer registry occupational information proved less successful than that reported by NIOSH's work (Weiss et al., 2015).

At the University of California, Los Angeles, the California Health Interview Survey is using NIOCCS coding in providing public search services to compare health demographic and insurance topics by industry and occupation. The new industry and occupation indicators in the California Health Interview Survey are coded with the help of the NIOCCS (UCLA, 2016). As described elsewhere, NIOSH is using NIOCCS to code BRFSS data on industry and occupation using interview records from 26 states. NIOSH is also starting new projects on coding cancer registry information from a small number of states and piloting use of real-time coding for death certificates in a sample of a few states. Building on the same knowledge base, NIOSH is adapting this approach for electronic health records to better serve clinician needs by preserving more of the rich details in occupation and industry titles with real-time coding.

BLS has also been developing software for coding occupational data in surveys, most notably the SOII. Previous research and ongoing monitoring of the SOII coding effort indicates that the software they developed is assigning codes as accurately or more accurately, on average, than their Office of Safety and Health Statistics' human coders.[3]

NIOSH and BLS have also been involved in coding research on other data elements relevant to surveillance: nature of injury, body part and event, result of injury, and source of injury codes (Measure, 2014; Bertke et al., 2016).

In addition to efforts by NIOSH and BLS to develop coding software, many academic groups around the world have also developed and evaluated tools for coding occupational data captured as free text from a variety of sources (Nanda et al., 2016; Marucci-Wellman et al., 2017). There does not appear to be adequate coordination of these activities across agencies and researchers.

**Conclusion: Several different coding systems are in use by agencies and other entities to record occupation and industry as well as injury and illness events. These systems have evolved to serve different objectives in different circumstances.** Surveillance of conditions that are attributed appropriately to work requires an effort to pursue coordination across agencies and other entities collecting relevant data. Examples include accurate and large-scale coding of occupational data of all types in EHRs, survey responses, death certificates, workers' compensation records, and related records that can be useful for surveillance of occupational conditions.

## ELECTRONIC EMPLOYER-BASED REPORTING OF OCCUPATIONAL INJURY AND ILLNESS

In 2016, OSHA issued a new electronic reporting rule requiring certain employers to submit establishment-level injury and illness data to OSHA (OSHA, 2017a). While the agency has required employers to keep injury

[3]BLS responses to the NAS OSH Surveillance Committee, August 19, 2016.

and illness records since 1971, for most employers, this information has only been available at the workplace, limiting the utility of the data. The new rule provides a much-enhanced source of injury and illnesses data that can be used for effective targeting of interventions and prevention efforts as well as compliance activity focused on hazardous industries, workplaces, exposures, and high-risk groups. Furthermore, these data are not currently available to agencies or the public from other surveys. This employer-based system also provides new opportunities to conduct outreach and build tools and provide assistance to employers to identify and address hazards at individual worksites.

For decades OSHA has utilized injury and illness data to help target its enforcement, compliance assistance, and other activities. With limited resources and staff—there are fewer than 2,000 federal and state OSHA inspectors responsible for overseeing the safety and health of over 140 million workers at nearly 8 million workplaces (AFL-CIO, 2016)—both federal OSHA and the state plans have sought to target their efforts on the most hazardous industries and employers and on the most serious and widespread hazards. But the agency's targeting and priority-setting efforts have been hindered by a lack of data, particularly establishment-level information, to evaluate the hazards and risks at individual worksites. While BLS gathers establishment-level injury and illness data for a representative sample of employers through the SOII, the information that is published or public includes summary estimates based on the experience of the sampled companies (weighted to be representative). No establishment-level data are made available to OSHA or the public, due to the BLS policy of maintaining confidentiality of the data in all surveys it conducts.

In 1995, OSHA launched its own initiative to collect injury and illness information from certain establishments. Under the initiative, called the OSHA Data Initiative, OSHA annually collected summary injury and illness information from approximately 80,000 establishments in selected high-hazard industries. This information was used to generate injury rates for individual establishments, with the establishments that showed the highest days-away-from-work injury rates placed on OSHA's site-specific targeting list for inspections. In more recent years OSHA made the data available through a search function on its website. The initiative was suspended in 2012 due to a combination of a reduced budget and OSHA's intention to replace it with expanded electronic reporting under the injury reporting rule.

In May 2016, OSHA issued a new rule requiring certain employers to report electronically injury and illness information required under OSHA recordkeeping regulations annually (29 CFR 1904) to OSHA. Under the new rule OSHA will be receiving establishment-specific injury and illness data from more than 460,000 worksites on an annual basis. The rule will provide injury counts and rates for all covered worksites and, for larger establishments, detailed case and demographic information on all injury cases, unlike the BLS SOII, which only collects detailed information on cases resulting in days away from work. OSHA plans to make much of the data collected from the electronic injury reports available on its website after scrubbing personally identifiable information and information restricted from disclosure under federal law. The Mine Safety and Health Administration (MSHA) has publicly posted injury and compliance results from all U.S. mines since the 1990s.

This new rule now requires electronic submission of relevant injury and illness reports that employers had already been required to maintain, but only onsite. Establishments with 20-249 employees in industries with historically high injury and illness rates will now be electronically submitting information from the OSHA Form 300A–Summary of Work-Related Injuries and Illnesses.[4] The electronic reporting rule covers more industries than were covered by the earlier OSHA Data Initiative, including many more industries in the service sector, which will provide valuable data for surveillance and intervention in this growing sector of the economy. For example, hospitals and ambulatory health care facilities, which both have high injury rates, are required to report injuries and illnesses to OSHA. All establishments with 250 or more employees that are covered by OSHA recordkeeping rules are required to submit information from the OSHA Form 300A, as well as the more detailed information being maintained already that is on the OSHA Form 300 (Log of Work-Related Injuries and Illnesses) and Form 301 (Injury and Illness Incident Report).[5,6]

The new rule will provide an extensive new data source regarding injury and illness that can be used by OSHA, NIOSH, state agencies, employers, workers, and researchers for a range of surveillance and prevention purposes. OSHA estimates that 34,000 larger establishments (≥250 employees) and 431,000 smaller establishments (with 20-249 employees) will provide information on the numbers of fatalities and injuries and illnesses

---

[4]Industries covered include agriculture, fishing, and forestry; utilities; construction; manufacturing; wholesale trade; and other industries with an average rate of days away from work, job transfer, or restriction of 2.0/100 employees or higher for 2011, 2012, and 2013 (OSHA, 2017b).

[5]The OSHA Form 300A provides an annual summary including information on the number of injury and illness cases, days away from work, and employment (numbers of employees and hours worked). The OSHA Form 300 provides a listing of each recordable injury and illness and includes information on the employee, the job/activity, the injury, incident, and days away from work or restricted activity. Form 301 is the incident report for each individual case and provides more detailed information on the employee, medical treatment, job activity, nature, source, and events and exposure related to the case.

[6]The new regulation also includes provisions to prohibit retaliation against workers who report injuries and policies and practices that discourage the reporting of injuries (29 CFR 1904.35 and 1904.36).

along with employment information that can be utilized to generate injury rates (OSHA, 2016). OSHA estimates that 34,000 logs and 700,000 injury incident reports will be submitted annually. As a benchmark, according to BLS, the SOII receives data from 240,000 establishments and data on 300,000 days-away-from-work cases (BLS, 2013).

The universe of establishments and cases covered by the OSHA and BLS collections are different. OSHA is collecting reports from *all* establishments with 20 or more employees in designated industries. For larger establishments (≥250 employees) OSHA will be collecting detailed data on *all* individual injury and illness case reports. BLS collects reports of a *sample* of establishments in each industry sector and a sample of case reports on days away from work cases in order to generate statistically valid estimates across all industry sectors. As described in Chapter 4, for all other injury cases (e.g., cases resulting in job restriction or job transfer), which represent 70 percent of all cases, BLS does not collect detailed case-level information in the SOII.

In addition, the information collected and available under the electronic reporting rule holds potential value for employers, workers, public health agencies, researchers, and others. Employers will be able to use the information to compare their experience with others in the industry. Workers will be able to have ready access to an employer's injury reports prior to seeking employment and while employed to assess the safety record of the employer. Public health agencies will be able to determine if there are types of injuries or illnesses occurring in the workplaces of particular industries. Public health departments will be able to initiate intervention efforts, including educational efforts and adjustments to public health standards in industries such as health care facilities, food establishments, or schools, which are regulated by the states. And researchers will have ready access to a large database of injury information to assist them with better characterizing high risks as well as assessing the effectiveness of interventions (O'Halloran et al., 2017).

The electronic reporting initiative also provides an opportunity to create a new avenue for expanding and targeting outreach to employers, particularly smaller employers, to assist them with hazard identification and prevention efforts. The agency could provide automatic feedback or reports to employers on how their injury rates compare with others in the industry. In addition, the agency would need to provide software and other tools and materials to employers to help them analyze their injury reports. Such feedback might be implemented as part of the Injury Tracking Application that OSHA is designing to collect occupational injury data directly from employers.

Among concerns raised about the new electronic reporting requirement, two are relevant to its potential role for occupational safety and health surveillance: duplication of reporting and questions regarding OSHA's capacity to utilize the substantial amount of new information it will be receiving. The OSHA electronic reporting rule and the BLS SOII could require some employers to submit the same information twice to the Department of Labor—once to OSHA and once to BLS. BLS estimated that there is a 40 percent overlap between the two reporting requirements (Monaco, 2016). OSHA and BLS are collaborating on the implementation of OSHA's electronic reporting rule so that BLS can use the data received by OSHA in the annual SOII. Such collaboration and coordination is critical to provide both agencies with the data they need while avoiding duplicate reporting requirements for employers. Furthermore, OSHA will have access to detailed data not available to the agency from the BLS-SOII efforts—data useful for prioritizing program efforts for targeting inspections and for efforts to support employers in compliance. The committee notes that currently there are some differences between the data included on the OSHA and BLS reporting forms that will need to be reconciled. Most specifically, the OSHA forms do not include race and ethnicity information for individual cases, which is an optional field on the BLS form.

Historically, OSHA's capacity to utilize data for enforcement and other purposes has been limited and concerns have been raised about the agency's ability to effectively utilize the data collected under this initiative. Similar concerns were noted in the 1987 NRC report when the issue of OSHA collection of employer injury and illness data was reviewed. As one response to the 1987 report, OSHA developed and implemented the OSHA Data Initiative, collecting summary injury data from 80,000 employers and utilizing the information for inspection targeting. As OSHA develops its system to receive and manage the electronic records, the agency will need to consult with BLS to assure that coding of the newly available case and demographic data that are submitted to OSHA as free text is compatible with the current methods that are in use for the BLS-SOII.

**Conclusion: The OSHA electronic reporting rule will serve a key role by providing data essential for injury and illness surveillance not available from the SOII.** These data are useful for targeting interventions and prevention efforts that focus on hazardous industries, workplaces, and exposures as well as high-risk groups. The rule also provides new opportunities to conduct outreach and to provide tools and assistance to employers who need to identify and address hazards at individual worksites.

Coordination and integration of data-collection efforts by OSHA and BLS will prevent duplication of reporting by some employers to both agencies which otherwise may undermine support for this new initiative. New data tools, including development of off-the-shelf software for use by employers or tools for OSHA to provide feedback directly to employers, will also be important in building support for this new initiative. Increased collaboration among OSHA, BLS, NIOSH, and state agencies will ensure the maximum use of this important new data source on work-related injuries and illnesses.

**Recommendation E: OSHA, in conjunction with BLS, NIOSH, state agencies, and other stakeholders, should develop plans to maximize the effectiveness and utility of OSHA's new electronic reporting initiative for surveillance.** These should include plans to provide ongoing analysis and dissemination of these data and to minimize duplication of reporting by employers.

In the near term:

- To avoid duplicate reporting, OSHA and BLS should integrate data-collection efforts so that employers selected in the annual BLS sample for SOII but reporting electronically to OSHA need not make separate reports to BLS. This will require that a unified reporting form include requiring race and ethnicity in submitted case reports.
- OSHA should provide timely and automatic feedback to employers that provides comparative information specific to the employer and others in that industry.
- OSHA should develop a publicly available and easily searchable injury and illness database based on the electronic reports.

In the longer term:

- OSHA and NIOSH should work with stakeholders to develop software and other tools and materials that facilitate further establishment-level analysis of injury data with specific attention to enabling effective use by employers as well as others to identify hazards and job-specific issues for prevention. With experience from participants in this electronic reporting, OSHA should explore feasibility to expand electronic reporting to all employers required to maintain OSHA logs.

## MOBILIZING USE OF WORKERS' COMPENSATION DATA

Historically, the state and national occupational health community has recognized the value of workers' compensation data despite their limitations (Utterback and Schnorr, 2010, 2013). For the most part, the data are useful at the state level where the workers' compensation claims data provide an extensive data source for case-based surveillance programs. Claims for specific conditions can be identified through queries of electronic databases, resulting in rapid and efficient case ascertainment. Claims data allow contact information for follow-back surveys to capture additional information about exposures, controls, and medical outcomes. Use of unique data sets within states that are the sole provider of workers' compensation insurance (exclusive state-funded states) has led to the characterization of conditions typically difficult to identify in other surveillance systems. For example, Washington State workers' compensation data allow characterization of specific work-related musculoskeletal disorders using ICD-9-based case definitions available in medical billing data (NRC and IOM, 2001; Silverstein et al., 2002; Marcum and Adams, 2017). Similarly, Liberty Mutual Insurance uses workers' compensation data to characterize the most common and costly types of workplace injuries, publishing the annual Liberty Mutual Workplace Safety Index (Marucci-Wellman et al., 2015). The California's workers' compensation information system similarly is building a comprehensive database of injury, lost-time cases, and medical care provision used in many system evaluation efforts and to promote targeting of prevention efforts (Das et al., 2012; Joe et al., 2014).

Workers' compensation data arise out of a system of oversight and case management. Data collected by state labor and workers' compensation agencies typically address four distinct phases of the workers' compensation case: the injury, the medical care, the financial compensation process, and the claims process. Injuries that become claims, particularly "lost-time" claims, beget much recordkeeping. A complete insurance record on a workers' compensation case includes information about the occupational injury or illness, about the provision of timely and adequate medical care, and about the payments for financial compensation for short- or long-term disability. A further record may be developed if there are disputes over how an injury occurred, whether it was on the job, the severity of the injury, or disputes over which employer holds responsibility.

Thus, workers' compensation records are uniquely positioned to collect data not generally provided by other injury and illness surveillance sources. These include data on injury and illness severity (e.g., number of office visits, hospitalizations, and days of compensated lost work time), direct costs of the components of medical care (e.g., office visits, procedures, medications, and physical rehabilitation), and financial benefits to the injured workers for time loss or permanent disability from work and/or benefits to their survivors. Furthermore, the collection of detailed medical billing data and medical records associated with care allows research on medical outcomes, and the appropriateness and quality of care (Prang et al., 2016).

Workers' compensation data are usually analyzed to determine how to price an insurance product for a single employer rather than using the data to focus on ways to lower the rate of injury and the associated social and economic costs that accompany occupational injuries. But in combination with other prevention-oriented surveillance tools, workers' compensation records hold promise to improve understanding and prevention of occupational injuries.

Due to limited coverage of illnesses that also occur in the general population (e.g., lung cancer), illnesses with long latency even when an occupational cause is likely (e.g., mesothelioma), and the difficulties associated with workers' compensation claims for illnesses, almost all workers' compensation claims are for injuries, musculoskeletal disorders, and acute illnesses. The potential use of workers' compensation data collated across states is also limited by different eligibility criteria in different states, changes over time in the medical management and work restriction policies of worker-related injuries, and varying data availability in different states (some states only computerize lost-work-time claims). Within a given state, the data have the potential to provide estimates of the magnitude of occupational injuries and select illnesses, trends in these conditions, emerging problems, and local variation in injury, and, when matched with other public health data, they can offer the expanded capacity to provide a more contextual and complete understanding of the landscape of occupational injury and some acute illnesses.

Researchers have increasingly used workers' compensation data for understanding injuries in specific occupations, industries, or trades, such as studies in law enforcement officers (Holloway-Beth et al., 2016), the trucking industry (Smith and Williams, 2014), and machine-related injuries in metal fabrication businesses (Yamin et al., 2016). This surveillance approach focusing on multiple types of injuries and health outcomes of an industry facilitates the development of prevention partnerships with affected unions and employers (e.g., TIRES, 2017).

The utility of workers' compensation data for public health surveillance, however, has been constrained for many reasons, among them administrative barriers to accessing data, data in unusable or burdensome formats for efficient use, and limited investment of public health resources to develop the technical expertise to use these data. Current advances in technology, electronic reporting and billing of workers' compensation claims data, development of standardized formats, and promotion of the value of workers' compensation data to the public health surveillance community are making workers' compensation data more desirable and usable for public health surveillance. Workers' compensation information is currently used in 3 of the 22 occupational health indicators recommended by the Council of State and Territorial Epidemiologists (CSTE, 2017). Workers' compensation data have also been used extensively in several states for surveillance of work-related injuries, including musculoskeletal disorders, as part of multisource surveillance (Kica and Rosenman, 2014; Largo and Rosenman, 2015) or where a state agency is the sole insurer for workers' compensation, most notably, Ohio and Washington.

In the United States, workers' compensation is primarily a state-based program[7] that provides no-fault medical care with no deductible or copay, and provides partial wage replacement (indemnity benefits) to workers injured on the job. The state systems vary markedly across the United States despite the recommendations for national minimum standards across state systems made in 1972 (National Commission on State Workmen's Compensation Laws, 1972). Each of the 50 states plus the District of Columbia has compensation systems, with varying eligibility requirements, benefit levels, dispute-resolution systems, and types of data that are collected for purposes other than claims administration. The U.S. Department of Labor ceased tracking the number of minimum standards met by each state in 1984. Such information is needed to explore using similar data from more than one state. For the most part,[8] these state programs are broadly based and mandatory for workers who are employees; however, they typically do not cover independent contractors, freelancers, or other self-employed persons. This distinction is especially important as increasing numbers of workers are considered part of the growing "gig economy." As an experience-rated insurance system, workers' compensation can theoretically provide incentives to employers to reduce their costs by making places of employment safer and healthier.

In many states, workers' compensation data are collected in a structured regulation-driven data system. Use of these data for direct national reporting and injury surveillance, however, is challenging. There is little uniformity among the states about what data and which transactions (and which versions of the transactions) are transmitted from claims administrators to the state. Some states mandate the reporting of claims using the Electronic Data Interchange standard[9] while others make reporting voluntary, while still others have no reporting rules to the state.

In addition to the inability to agree on a set of mandatory national

[7]There are also several federal programs covering specific categories of workers: federal workers (Federal Employees' Compensation Act); longshore and harbor workers employed in maritime employment upon the navigable waters of the United States; coal miners who qualify for programs to compensate for black lung disease (miners totally disabled by pneumoconiosis arising out of coal mine employment), and to survivors of coal miners when deaths are attributable to the disease; and current or former employees (or their survivors) of the Department of Energy, its predecessor agencies, and certain of its vendors, contractors and subcontractors, who were diagnosed with a radiogenic cancer, chronic beryllium disease, beryllium sensitivity, or chronic silicosis, as a result of exposure to radiation, beryllium, or silica while employed at covered facilities (the Energy Employees Occupational Illness Compensation Program Act).

[8]Texas allows employers to "opt out" of mandatory workers' compensation coverage.

[9]International Association of Industrial Accident Boards and Commissions' Electronic Data Interchange Claims Standards "are used by claims administrators to report workers' compensation first report of injury and subsequent report of injury claims data to U.S. jurisdictions" (IAIABC, 2017).

standards, state workers' compensation agencies have proved unwilling (and maybe unable) to standardize across systems, share data with other jurisdictions, or share with any central body. In most states, in contrast to voluntarily collected data under the BLS SOII system, there is no collection of race and ethnicity information. Other challenges to using workers' compensation data in national surveillance efforts are challenges to comparability: differing scopes of coverage or varying waiting period before temporary total disability benefit receipt begins or when injuries need to be reported to administrative agencies.

A different type of barrier to the use of workers' compensation data is underreporting of work-related injuries and recognized occupational illnesses to these systems (Rosenman et al., 2000; Azaroff et al., 2002; Fan et al., 2006). Reasons for underreporting are numerous, including worker and health care provider's limited understanding of the available worker compensation benefits due an injured worker, characteristics and severity of the injury or illness, the administrative burden filing a claim imposes upon the worker and health care providers, availability and access to alternative health care benefits, individual worker characteristics, the workers' employment relationships to the workplace, and disincentives placed upon the health care provider and worker by the employer to claim workers' compensation benefits.

Variation exists in the underreporting of worker-related injury and illness, benefit eligibility restrictions imposed by varying state statutes, and regulations and case law across U.S. states. This limits the value of workers' compensation data for interstate comparisons of occupational injury and illness rates (Bonauto et al., 2010). The potential variation in underreporting based on individual and employment characteristics may limit the value of workers' compensation as a singular source of data for establishing safety and health priorities within a state (Fan et al., 2006).

Even with their limitations, workers' compensation insurance data include unique information not available such as medical and wage replacement cost data and information on medical care provided as well as work outcomes. Cost data can be categorized and reported in conjunction with claim incidence rates to help target high hazard and high economic cost industry groups (Bonauto et al., 2006). Medical and disability outcome data provide opportunities for surveillance and research. Developing measures to track the quality of care provided to injured workers (e.g., use of opioids for noncancer pain in workers' compensation) or measures to assess the timeliness of care in the treatment of common occupational conditions are unexplored areas for the occupational health surveillance community. In Australia, a Compensation Research Database is being assembled to study the influence of compensation system processes and practices on health and health-related outcomes (Prang et al., 2016).

While most federal systems use standard industry and occupation classification systems (i.e., NAICS for industry and the SOC system for occupation; or Bureau of Census Classification for industry and occupation), workers' compensation offers data classification systems that supplement these traditional classification systems (Spector et al., 2014). The most common are "risk classes" or "manual class codes" (see earlier discussion) that group workers based on similar risk for financial loss. Job tasks within an occupation or industry vary, and risk classes by design attempt to capture these differences. For example, employees in a grocery store are cashiers, butchers, and delivery drivers, all of whom would be identified by a common establishment industry code but each would be separated in workers' compensation data with individual class codes assigned. Strategic use of manual class codes may be useful to identify specific high-hazard job tasks occurring within an otherwise seemingly low-risk industry or occupation.

### Opportunities for Incorporating Insurance Data and Improving Workers' Compensation Data for Occupational Safety and Health Surveillance

Insurers can both contribute to and benefit from the increased amount of risk information that comes from improved surveillance. Insurers can combine their own information on site-specific hazards from safety engineering and industrial hygiene resources within their companies, with improved information from broader industry focused exposure surveillance and industry-wide injury statistics. The aggregation of claims information in risk classifications by national, regional, and state specific insurance rating organizations is used to identify emerging hazards and outcomes for companies purchasing workers' compensation coverage. In the past, in some specific industries and workplaces, workers' compensation and health insurance information has been combined to indicate types of injuries and illnesses that would not otherwise become apparent in workers' compensation claims alone. Re-insurers are in good position to look more globally at such emerging risks as they might take on coverage of industrial hazards that are not apparent to individual insurers.

The facilitation of claims data sharing among state workers' compensation systems, OSHA programs, workers' compensation insurers, and health departments may be useful in targeting scarce governmental and private resources for health and safety, including both governmental enforcement and consultation services, and insurer loss control functions. Surveillance findings could also be disseminated broadly through workers' compensation insurers to their insured employer clients as they emerge. The NIOSH Center for Workers' Compensation Studies (described below) can be an important bridge between these different entities.

A few methodological research areas could help improve the use of workers' compensation data for OSH surveillance. BLS and NIOSH have engaged in research toward coding of both injury narratives and text fields that document occupation and industry. This work could also contribute to crosswalks between distinct systems in the standardized OSH surveillance systems and workers' compensation categorization methods used in the insurance industry, such as by the NCCI and the International Association of Industrial Accidents Boards and Commissions.

In 2012, NIOSH established the Center for Workers' Compensation Studies (CWCS) (Utterback and Schnorr, 2013). According to NIOSH, this center focuses on three areas: "(1) expanding use of state-level workers' compensation claims data for research and prevention, (2) identifying and communicating interventions most effective at preventing injury and illness, and (3) encouraging collaborations between the public health and workers' compensation communities" (NIOSH, 2016c). The program has begun to "build the capacity of states to use workers' compensation claims data for prevention purposes through grants, partnerships, and technical assistance" (NIOSH, 2016c). It also "evaluates approaches to preventing illness and injury by working with workers' compensation insurers" (NIOSH, 2016c). Through information distributed to insurers, state workers' compensation bureaus, and state departments of health, the center has focused on best practices for treatment of illness and injury, data analysis and denominator considerations, and issues related to return to work. The center is currently working with five states—Ohio, California, Massachusetts, Michigan, and Tennessee—to use claims data toward focusing efforts of injury prevention in high-risk industries (Wurzelbacher et al., 2016).

The center's orientation is particularly important given the lack of injury prevention–oriented research in most state workers' compensation agencies. The center is working to bring state public health epidemiologists associated with OSH surveillance programs together with workers' compensation data experts to focus the data on primary, secondary, and tertiary prevention of work-related injury and illness.

**Conclusion: Workers' compensation data have potential value for surveillance of occupational injuries at the state level and have been specifically used to this end in three states (Michigan, Ohio, and Washington).** In 2015, NIOSH established the CWCS to promote use of workers' compensation data to improve workplace safety and health. The Center works to develop new methods for coding, analyzing, and disseminating workers' compensation data, foster new research collaborations, and share best surveillance and research practices across state agencies, researchers, and insurance companies using these data. Because workers' compensation laws and eligibility requirements vary by state, surveillance findings based on workers' compensation records cannot be readily compared across states. The Department of Labor no longer tracks how well each state meets minimum standards. Surveillance use would be enhanced if this tracking were resumed and made public, and possibly new minimum standards would apply to all states considered.

**Recommendation F: NIOSH with assistance from OSHA should explore and promote the expanded use of workers' compensation data for occupational injury and illness surveillance and the development of surveillance for consequences of injury and illness outcomes, including return to work and disability.**

In the near term:

- NIOSH should organize an advisory group of workers' compensation data experts to advise both the NIOSH Center for Workers' Compensation Studies and interested states concerning their use of workers' compensation data for surveillance and research.
- NIOSH should encourage states to expand the use of workers' compensation information beyond the Council of States and Territorial Epidemiologists (CSTE) occupational health indicators. Specifically, the agency should work through the state surveillance cooperative agreements to develop and enhance use of workers' compensation data for state-based occupational injury and illness surveillance and prevention activities.

In the longer term:

- NIOSH and OSHA should collaborate with states to pursue the development of surveillance systems that capture cost of work-related injury and illness, measure work-related disability and return-to-work outcomes, and assess the adequacy of benefits administered through workers' compensation insurance programs.

## LEVERAGING EXISTING SURVEYS AND DATA SYSTEMS

NIOSH's surveillance strategy places importance on leveraging existing surveys and data systems. Such efforts provide a valuable addition to the limited resources that NIOSH has available to dedicate specifically to occupational safety and health surveillance. In some examples, such as the National Health Interview Survey described in Chapter 4, additional questions concerning work-relatedness of conditions, common workplace exposures, and other issues related to work have been incorporated in the survey. NEISS-Work, also described in Chapter 4, is another example of collecting more extensive information about work and work-relatedness in an existing surveillance system. Listed in Table 6-1 are some examples

**TABLE 6-1** Examples of Surveys and Studies Leveraged to Generate Occupational Health and Safety Information

| Survey | OSH Objective | NIOSH Engagement |
|---|---|---|
| National Birth Defects Prevention Study (NBDPS) | Examine potential associations between role of work (parental jobs and workplace exposures) and birth defects and other pregnancy outcomes (Lupo et al., 2012). | Partnering with the NBDPS to examine the role of work on birth defects and other pregnancy outcomes. From this study, associations have been found between several parental jobs and workplace exposures and specific birth defects. |
| National Ambulatory Medical Care Survey Asthma Supplement | Learn more about work-related asthma management strategies and track improvements. | Examine work-related asthma management strategies to measure barriers to work-related asthma management strategies, and acceptance of the guidelines by health care practitioners. |
| National Crime Victimization Survey (NCVS) | Provide annual information on burden and track changes in patterns of work-related violence. | Improve data on work-related violence collected through the NCVS. |
| National Health and Nutrition Examination Survey (NHANES) | Examine and track work associations with documented health conditions—initial emphasis on respiratory conditions using spirometry results. | Provide the technical support to ensure high-quality spirometry data to study obstructive lung disease and to collect data on workplace exposure for occupational surveillance purposes. |
| Fatality Analysis Reporting System (FARS) | Measure and track work-related motor vehicle accident fatalities. | Collaborate with the National Highway Traffic Safety Administration and BLS to link CFOI and FARS data for comprehensive data on fatal occupational crashes across industries and vehicle types. |

of national surveys that NIOSH is taking advantage of, as opportunity and resources allow, to incorporate additional questions and generate new occupational safety and health information. An important advantage of collecting information about occupational health and safety within broader public health data sources is that, unlike employer-based reporting systems, this approach allows for assessment of the contribution of work to the overall problem under investigation (e.g., asthma prevalence and incidence of violence). It also offers opportunities to collaborate with other public health programs to develop more comprehensive approaches to understanding and addressing multifactorial public health problems. These efforts suggest that NIOSH, working in collaboration with other agencies, could take further advantage of existing surveys and data sources to go beyond filling information gaps in order to help meet specific surveillance objectives.

There are, however, significant challenges in getting multiple work-related questions into existing surveys and studies that have been designed for other purposes. These include competing public health or other agency priorities and increasing pressure to limit lengths of surveys to address declining response rates and to reduce costs. Failure to recognize the importance of assessing the impact of work on health within the broader public health community tasked with addressing other public health problems is also common. Ultimately, final decisions about survey content are beyond the control of NIOSH or state occupational health programs and continued (ongoing or periodic) data collection is accordingly unpredictable.

NIOSH is also collaborating with other federal agencies and state partners to promote routine collection and coding of basic information about industry and occupation in existing health surveys and other public health surveillance systems. Collection of this information not only increases potential use of these data sources for OSH surveillance and research, but in many instances, it can also enhance practice in other public health domains by providing information about patterns of health outcomes and determinants of health (e.g., prevalence of smoking behaviors and access to preventive services, and access to health insurance) in relation to work. Given the many industry and occupation categories, this information is generally collected in surveys as narrative text, unlike most other survey variables, which imposes a substantial coding burden. Automated approaches to assigning standardized industry and occupation codes to narrative text are therefore essential to gain acceptance for collecting these data elements and realizing the potential of utilizing the existing surveys and data systems (see above section on coding and further discussion in Chapter 7). Advances in electronic coding of industry and occupation described elsewhere in this report demonstrate much promise in meeting this challenge and have facilitated some successes described below. Further support for maintaining and enhancing these tools is needed.

Occupation and industry are core sociodemographic variables collected in the decennial census and other population and economic surveys including, among others, the Current Population Survey and the American Community Survey. Given the connectivity of one's occupation to an individual's level of education, it is also an important component that impacts an individual's economic status. The importance of documenting health status in relation to type of employment has a long history in the United States. In 1847, Massachusetts was the first state to establish a death registration system including information about occupation on the state death certificate, with subsequent annual reports on mortality by occupation. Today work is widely recognized as an important social determinant of health, having both direct impacts though the physical and psychosocial environment as well as indirect impacts through access to economic and health resources (Wilkinson and Marmot, 2003; An et al., 2011). Information about industry and occupation are currently collected and coded in most federal health surveys, including NHIS, NHANES, the Medical Expenditure Panel Survey (MEPS), and the National Survey on Drug Use and Health, which gathers information about substance use and dependence or abuse. Industry and occupation are particularly salient sociodemographic measures in national health care surveys such as the MEPS when determining an individual's access to health insurance and the comprehensiveness of the coverage, given the primary source of coverage for a substantial representation of the population is employment based.

In light of the importance of the inclusion of these measures in such health-related surveys, an impending analytic challenge may develop as a consequence of the current revisions under way in NHIS which is considering collecting industry and occupation information only on a rotating basis (NCHS, 2016). Furthermore, there are also some relevant nationwide health surveys and many CDC public health surveillance systems in which these data elements are not collected, thus limiting the usefulness of these systems for identifying potential cases of work-related disease and characterizing patterns of health outcomes under surveillance in relation to occupation and industry. In some instances, such as CDC's Pregnancy Risk Assessment Monitoring System (PRAMS), the employment information (maternal occupation and industry) is not collected at all although it is collected on birth certificates in some states.[10] There are other information sources, such as state and local cancer registries, where the industry and occupation data are collected and maintained (when present in the medical records) at the state level, but they are not routinely coded and not reported centrally to the National Program of Cancer Registries at CDC which funds state cancer registries in 45 states and uses aggregated state data to generate the official federal statistics on cancer incidence.

NIOSH has had some success in recent years working with other CDC programs and state partners to promote collection of industry and occupation information in additional data sources used for public health surveillance. A major critical initiative is NIOSH's ongoing work to incorporate industry and occupation information in electronic health records (described

[10]The CDC PRAMS survey, conducted in partnership with state health departments, collects information annually on approximately 83 percent of all U.S. births specifically regarding experiences and behaviors before, during, and soon after pregnancy (CDC, 2017b).

above). NIOSH has under consideration a pilot effort with the National Center for Health Statistics to code industry and occupation data on death certificates from 17 states in real time, which, if successful, will provide the opportunity to extend the effort and analyze mortality patterns by industry and occupation for all 50 states.

**BOX 6-4**
**Example: Collection of Industry and Occupation Information in the Behavioral Risk Factor Surveillance System (BRFSS)**

BRFSS is a national continuous health survey, conducted by telephone, of a representative sample of U.S. adults, administered in all states as a collaboration between the CDC and state health departments. Data are collected on a variety of health risk factors, preventive behaviors, chronic conditions, and emerging public health issues. In addition to national estimates BRFSS provides information to target prevention and monitor progress in meeting prevention goals at the state level. All states implement a set of core questions and can choose to implement optional national modules on various topics. For example, an ongoing national asthma module collects information specifically about work-related asthma and has been used to estimate the proportion of adult asthma attributable to workplace exposures and the extent to which health care providers ask patients with current asthma whether their asthma is work related (a Healthy People 2020 objective). States may also incorporate their own modules to address state-specific concerns.

Information about industry and occupation is not routinely collected in the national core BRFSS module. NIOSH is currently supporting collection of these data elements in the survey on a pilot basis in 26 states with the aim of encouraging inclusion of this information as core demographic variables in the future. NIOSH codes the industry and occupation data for participating states using the automated NIOCCS (see above) with computer-assisted coding by trained coding staff. This work has demonstrated the feasibility of collecting and coding industry and occupation information in the BRFSS. NIOSH has published findings based on data aggregated across states and states have published state specific findings. (Examples of recent publications using the industry and occupation data in the BRFSS include NM DH, 2014; NH HHS, 2015; Towle et al., 2015; UT DH, 2015; CDC, 2016a,b; LA DH, 2016; MA DPH, 2016; VT DH, 2016; WA DLI, 2016; O'Halloran et al., 2017; Shockey and Wheaton, 2017.)

Other CDC centers and public health programs at the state level have requested BRFSS data sets with occupational information for analyses or collaborated in conducting analyses. A continuing challenge is the length of the survey. In 2011, 2015, and again in 2017, NIOSH proposed that industry and occupation be added to the core BFRSS. While the majority of states have been in favor, the required 70 percent of states[1] did not affirm inclusion in 2011 and 2015 (the outcome in 2017 is still pending) primarily due to concerns that the survey core is already too long and other questions need to be deleted before adding more.

---

[1]Questions have to be approved by 70 percent of state BRFSS coordinators.

NIOSH is currently compiling and coding the industry and occupation data from six state cancer registries and will conduct analysis to assess associations between cancer incidence and industry and occupation in aggregate and stratified by state. Additionally, NIOSH has provided support to pilot the collection of maternal industry and occupation in the PRAMS survey in five states. Box 6-4 describes another initiative working with the states on the Behavioral Risk Factor Surveillance System that is illustrative of both the opportunities and the challenges in incorporating occupational information in public health data systems.

The CSTE has recommended that "occupational and industry and other work information as appropriate be included within CDC surveillance systems where feasible" (CSTE, 2014b). Also, the National Committee on Vital and Health Statistics' Subcommittee on Population Health, charged with recommending minimum standards for measures of socioeconomic status for federal health surveys, has recommended to the Department of Health and Human Services (HHS) that occupation and industry be collected as socioeconomic variables (also referred to as demographic variables) in all federal health surveys (NCVHS, 2012).

For public health surveillance systems that rely on data from health care providers and administrative data from the health care system, lack of information about occupation and industry in medical records is an underlying obstacle. Concurrent efforts are therefore needed to promote collection of industry and occupation in EHRs, as well as to develop automated approaches to coding industry and occupation (see above recommendations) to realize a 21st-century vision in which industry and occupation information is routinely collected and coded in relevant public health surveillance systems.

**Conclusion: Occupation and industry are demographic variables that describe core features relevant to adults and are characteristics of individuals essential to understand fully health and the factors that influence it.** These variables are considered core demographic variables in the decennial census but are not currently treated as such in systems that collect information on health. Public health information on health and disease among the adult working age population is gathered through surveys or collected in data systems primarily located in the Centers for Disease Control and Prevention's National Center for Health Statistics and in the Agency for Healthcare Research and Quality (AHRQ). On occasion these have included information on occupation and industry that allow characterization of health by work characteristics. The inclusion of occupation and industry information as a core demographic variable in systems designed to inform the nation about adult health would add important information to guide disease and injury prevention and delivery of health care in the population.

**Recommendation G: HHS should designate industry and occupation as core demographic variables collected in federal health surveys, as well as in other relevant public health surveillance systems, and foster collaboration between NIOSH and other CDC centers in maximizing the surveillance benefits of including industry and occupation in these surveys and surveillance systems.**

In the near term:

- HHS should re-establish industry and occupation as core demographic variables in all federal health surveys.
- CDC surveillance programs, as they proceed with their state partners to streamline and harmonize data across systems, should work with NIOSH to identify appropriate processes for collecting and coding occupational and industry data.
- NIOSH with assistance from CDC should explore and prioritize public health surveys that can be used to enhance occupational health surveillance objectives by collecting relevant occupational information.

In the longer term:

- To promote proper analysis of surveillance data NIOSH should develop methods and training materials on approaches to basic as well as new and creative use of occupation and industry data and on the selection and use of appropriate labor force denominators.

## IMPROVING OCCUPATIONAL HAZARD AND EXPOSURE SURVEILLANCE

A key component that is largely missing from U.S. occupational safety and health surveillance is collection and analysis of data on occupational hazards and exposures. Hazard and exposure data are leading indicators for anticipating and preventing work-related chronic disease and, to a lesser extent, acute disease and injuries. As noted in Chapter 4, useful information about hazards and hazardous exposures requires the identification of the hazard and the assessment of the parameters regarding the exposure (i.e., duration and intensity).

Hazard and exposure surveillance provides several benefits not achievable in surveillance of health outcomes, at least for certain classes of occupational risk. First, and perhaps most obvious, is that exposure must precede an adverse outcome, thus, at least in theory, supporting the timely mitigation of the risk prior to injury or illness. For most conditions of interest, there are irreversible consequences, from lost productivity and income due to days lost at work due to an injury, to development of a chronic disabling disease or death; effective intervention requires assessing the risk prior to the effects occurring. While disease surveillance may be effectively used to prevent future harm, only exposure or hazard surveillance fulfills the primary prevention of occupational ill health.

Related to the timeliness of hazard and exposure surveillance is the opportunity to provide timely feedback to employers interested in maintaining the effectiveness of exposure controls—both for prevention purposes and to ensure compliance with government regulations. As discussed in Box 6-5 a silica exposure surveillance scheme has demonstrated the effectiveness of rapid feedback and benchmarking of exposure monitoring results in further reducing workplace exposures to silica. Collection of such routine exposure data can be integrated into an employer's occupational health program to ensure compliance with regulatory limits, as well as being part of a government agency's routine collection of compliance information. Thus, a surveillance system can be built as part of other programmatic goals and does not have to be solely dedicated to surveillance.

The second compelling argument for exposure surveillance is in the context of multifactorial chronic diseases, such as cardiovascular disease, obesity, chronic lung disease, and musculoskeletal disorders. For those diseases with established exposure-response relationships, the availability

**BOX 6-5**
**Industry-Based Exposure Surveillance: An Example from the European Industrial Minerals Association Dust Monitoring Program**

In 2000, the European Industrial Minerals Association established a comprehensive exposure surveillance program for silica exposures among their members. The key features of the program were

- a defined exposure assessment strategy and measurement protocol,
- the centralization of results at an independent university-based institute, and
- the timely reporting of results back to the individual enterprises.

By 2015, 160 worksites in 23 countries had reported almost 28,000 samples (Zilaout et al., 2017). Most importantly, the results show a substantial downward trend in respirable crystalline silica exposures, especially during the initial years of participation. This project demonstrates the potential impact of a well-constructed exposure surveillance system for risk reduction and the potential for private industry to participate and benefit from this activity.

of exposure data allows for identification of workplaces with excess risk, and thus opportunities to reduce the occurrence of these multifactorial conditions.

On a population basis, the attributable fraction of a specific condition can be derived based on a comparison of risk in an "exposed" population compared to an unexposed control population. Such comparisons are usually made based on work in a specific industry and/or occupation and are not well informed by the actual probability and level of exposure within those populations. Thus, even this population-level estimate is based on exposure assumptions and is usually fraught with misclassification error. The error can be controlled only to the degree that specific exposure information is available on a population.

A compelling example of using the attributable risk to conduct risk assessment based on the population distribution of exposure is illustrated in the estimation of the future burden of work-related cancer in the United Kingdom in Box 6-6. This powerful approach to understanding the risk of chronic multifactorial disease is the key strength of a robust exposure surveillance system.

The uses of a robust exposure surveillance system are many. The primary use is, of course, the rapid identification of emerging issues or trends in risks and the use of such information for workplace interventions. The collection and analysis of exposure or hazard data, however, can also be used for epidemiological studies, both identifying new risks and refining the quantification of risk for those already identified. Exposure data are also instrumental in conducting risk assessments to understand how and where population risks of disease will occur, estimating the cost burden associated with such risks. Finally, studies based on good quantitative exposure data are needed to formulate policies to prevent or mitigate health impacts and the costs associated with alternative policy choices.

## Approaches to Hazard and Exposure Surveillance

Several approaches to hazard and exposure surveillance (described below) have been or are currently under way through the work of NIOSH, OSHA, and international organizations. Learning from those experiences and building on them will set in place a systematic approach to hazard and exposure surveillance with the potential to greatly improve worker safety and health. Data for such surveillance come from a wide variety of sources. Figure 6-1 shows a Venn diagram of the many sources of data and the overlap among the systems that collect these data for occupational exposure surveillance. It provides an overview of the roles and relationships for many of the systems further discussed in this chapter.

As with injury and illness outcomes, exposure information is derived from multiple sources, including individual, public agency, or employer. Figure 6-1 illustrates how these various exposure data sources relate to each other. While the three circles representing public agencies, individuals, and employers each have multiple sources within them, the totality of the data derived from the three major types of exposure information also may intersect. *Public Agencies:* OSHA, its state-based affiliates, and MSHA collect exposure monitoring data during regulatory or consultation inspections. NIOSH, through Health Hazard Evaluations and other research-oriented studies, collect substantial exposure monitoring data. Not included but potentially relevant is the hazard information contained in the Department of Labor's Occupational Network Database (O*NET) and Occupational Requirements Survey (ORS). *Employers:* Employers, or consultants working on their behalf, routinely monitor exposures among their workforce,

**BOX 6-6**
**Example of Exposure Data Used to Estimate Burden of Chronic Disease**

To estimate the incident cases and fatalities, Rushton and colleagues (2012) in the United Kingdom have identified well-established exposure-response studies for work-specific cancer risks. In this effort, the impact of all known human carcinogens, based on IARC classifications, was estimated, including well-recognized agents such as asbestos, silica, polycyclic aromatic hydrocarbons, etc., but also shift work, sunlight exposure, etc. However, to estimate the burden of disease in the population posed by these hazards, it was also necessary to estimate the prevalence and intensity of exposure to each among the working population in Great Britain.

Historical levels of employment were combined with a comprehensive exposure matrix, CAREX, to estimate the distribution of exposures to each agent over time (van Tongeren et al., 2012). As a result of this extensive exercise, the researchers were able to estimate that about 5 percent, or more than 8,000, cancer deaths in the United Kingdom in 2005 were caused by known occupational carcinogen exposures (Rushton et al., 2012). Furthermore, the models developed could be used to predict future burden of occupational cancers under current levels of exposure and inform alternative policy decisions which could affect the frequency or intensity of exposure, thus making invaluable contributions for health and economic planners and policy makers (Hutchings et al., 2012). The ability to provide these estimates is largely dependent on having population estimates of occupational exposure, including prevalence in the population, and their duration and intensity. By far the greatest limitation of the estimates produced is the lack of up-to-date quantitation of the exposure parameters. Similar studies have now also been conducted in Canada and Australia.

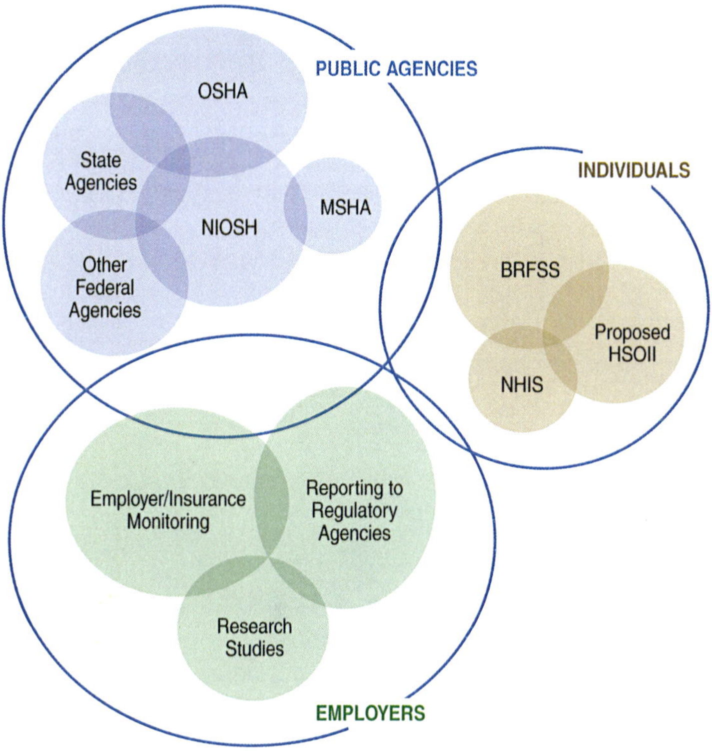


**FIGURE 6-1** Current Possible Overlaps in Data Sharing from Major Sources of Data Used for Occupational Exposure Surveillance. The size of inner circles do not represent relative importance of the source.

though primarily for regulated agents. A small fraction of the monitoring done by employers may also be reported to the agencies regulating them. *Individuals:* While individuals generally do not conduct exposure monitoring, they contribute important exposure-related information through various population based surveys such as the BRFSS and NHIS. Not included is the self-reported hazard information in the NSF-NIOSH Quality of Worklife Survey (QWS). Expansion on the types of exposure information collected using the Household Survey of Occupational Injuries and Illnesses could be integrated with exposure measurement databases, and thus the overlap of the individual and public agency data sources. Also missing from this diagram is one additional source of exposure information: biological monitoring results which may be collected through employers or other means and collected by analytical laboratories.

**Workplace-Based Direct Observation Surveys**

From 1972 to 1974 NIOSH conducted the National Occupational Hazard Survey (NOHS) in approximately 5,000 workplaces (Frazier, 1983). The surveys included a walk-through inspection in which engineers were supposed to observe "every plant process and every employee," making estimates of the numbers of workers exposed full time and part time to various substances and collecting other information, including data on whether engineering controls had been implemented or personal protective equipment was required for specific exposures. A similar survey, the National Occupational Exposure Survey (NOES), was conducted by NIOSH from 1981 to 1983. Neither of these surveys included any industrial hygiene sampling. Despite the ground-breaking utility of these surveys at the time, no follow-up surveys of a similar nature have been conducted since, largely because of the expense.

Direct observation of exposure conditions within a representative sample of workplaces offers benefits that include

- Characterization of both the use and exposure potential of multiple hazards at the same time, while also characterizing the number of individuals or probability of exposure occurring to the workforce.
- Obtaining a distribution of the frequency, intensity, and duration of such exposures, thus providing a complete distribution of both hazard and exposure, and allowing direct calculation of risks.

However, some risks are not easily observed, and within any workplace there may be too many hazards for an individual observer to identify as many risks may occur in highly incidental conditions and be difficult to observe on any specific day. Selection of a sample of representative workplaces is increasingly difficult due to the mobile and temporary nature of work organization in many industries. The biggest challenge for this approach is its complexity and expense.

*Exposure Measurement Databases*

Quantitative measurement of exposure intensity using personal sampling methods provides the "gold standard" for industrial hygiene monitoring and standards compliance and provides the most accurate means of evaluating current exposure levels. Because most regulatory compliance is based on quantitative exposure measurement, both industry and government agencies collect a large number of exposure measurements in a wide range of industries. There is powerful potential for compiling these routinely collected quantitative measurements into databases, which could then be used for surveillance activities.

For instance, the compilation of MSHA data on silica or coal dust during various mining activities has allowed for active surveillance of conditions and risks, intervention at locations with high exposures, and for epidemiology studies of dust-related health conditions. Additionally, OSHA includes quantitative industrial hygiene measurements in their publicly available Integrated Management Information System (IMIS). Inspector-collected samples that were analyzed in OSHA's Salt Lake City laboratory were included in the Chemical Exposure Health Data (CEHD), made publicly available since 2010 as part of the OSHA Information System. At present the CEHD contains quantitative industrial hygiene measurements and related information on the analyses for Salt Lake City laboratory samples only. The CEHD and IMIS have a significant degree of overlap (about 50 percent) but each data set contains a substantial amount of unique data and the IMIS may include an underrepresentation of nondetectable results (Lavoue et al., 2013a,b). The IMIS database has provided the opportunity for monitoring exposure levels of a few key agents in the workplace throughout the United States. However, the limitations of such data and the challenges in collecting all the relevant information (e.g., the distribution of those exposures among the working population, the frequency of their encounter, the duration of their occurrence, and the wide range of potential agents to be measured) need to be recognized.

### *Biomonitoring*

Validated biomonitoring methods have been developed for some exposures, such as lead in blood, metabolites of selected pesticides in urine, and cadmium in urine. Biomonitoring has the advantage of circumventing limitations of environmental monitoring, such as the effectiveness of personal protective equipment used by the worker, but has several challenges including potential invasion of privacy and sensitivity to the timing of the sample. Nevertheless, blood lead surveillance, both based on OSHA requirements for monitoring for lead-exposed workers and reporting of blood lead results by analytic laboratories, has proven highly effective in identifying high-risk conditions and stimulating personal and workplace interventions where needed (see Chapter 4).

### *Job Exposure Matrices*

Classical job exposure matrices (JEMs) were developed as relatively crude and static matrices of jobs with assigned exposures, usually nominal exposure based on self-report or expert judgment; however, JEM methodology has continued to be refined to allow for incorporation of quantitative measurements, refined job categories, information on the duration of exposure, etc. Two examples are worth noting: FINJEM, a comprehensive JEM based on Finnish industry information that allows for periodic updates and includes quantitative measurements (Kauppinen et al., 2014), and CAREX, focused on workplace carcinogens and first developed to cover all European Union member states (Kauppinen et al., 2000) and later tailored to Canada and numerous other countries (also see Chapter 5). Both systems may be used for exposure estimation of populations where an independent data set for duration within each job category and time period is available. Estimates derived in this way are useful for surveillance (see Chapter 5, Figure 5-1 and Box 6-6) and conducting epidemiological or risk-assessment studies for chronic disease. These JEMs include physical agents, chemical agents, biologic agents, physiologic and ergonomic factors, as well as psychosocial factors. As a consequence, the disease risks that can be monitored range from cancer to mental illnesses.

The use of job exposure matrices in surveillance activities could be limited by the extent to which the matrix information can be updated to reflect current conditions. However, the combination of a well-established population-specific JEM with ongoing collection and integration of exposure measurements could prove a powerful surveillance approach.

Given the importance of exposure duration in the estimation of risk, it is vital that sources of information about time spent in various exposure scenarios are also addressed. Questionnaires and administrative data can both be used to estimate risk, assuming the categories of activity addressed can be linked in some way to exposures. Such data may be linkable to exposure level using a JEM, for instance.

### *Other Sources of Generic Job Exposure*

O*NET is a publicly available online database that describes occupational features across U.S. job titles (O*NET, 2017). The database is continually updated by surveying a broad range of workers from each occupation. O*NET has been used to estimate workplace physical and psychosocial exposures and organizational characteristics. The data from a generic job exposure system offers the potential to impute hazard presence associated with a job. This, in turn would allow identification and tracking of potential occupational risks that otherwise have escaped consideration due to missing data or resource constraints on direct collection of job exposure information in the field (Cifuentes et al., 2010). For example, Gardner and colleagues (2010) noted that "job title–based exposure estimates from O*NET and self-reported and observer-rated exposures showed moderate to good levels of agreement for some upper extremity exposures, including lifting, forceful grip, use of vibrating tools, and wrist bending" (Cifuentes et al., 2007; Evanoff et al., 2014). O*NET also provides job information that could be useful to evaluate psychosocial working conditions especially for the demand/control and effort/reward models (Cifuentes et al., 2007; d'Errico et al., 2007; Boyer et al., 2009; Meyer et al., 2011). Further validation of these data is necessary to determine the utility of the O*NET databases for surveillance.

While O*NET has been partially validated, the Bureau of Labor Statistics, National Compensation Survey (NCS) program has initiated a refined "Occupational Requirements Survey" (ORS) with additional detail about the prevalence and level of physical demands and environmental exposures (BLS, 2014b, 2017b). While designed for the Social Security Administration (SSA) to describe occupation requirements that will assist the agency in eligibility determinations for Social Security Disability Insurance and Supplemental Security Income disability benefits for applicants, the ORS may prove a more useful source for JEM information for some prevalent exposures. For example, this database may supplement the O*NET with a more nuanced description of some of the cognitive and mental requirements for a job as well as adding information about the duration of specific physical demands and environmental exposures associated with jobs. (See the discussion on challenges of using survey data for exposure assessment in the next section on questionnaire surveys.)

### *Questionnaire Surveys*

Questionnaire surveys of working populations are another powerful and underappreciated approach to exposure surveillance. Survey responses are well suited to characterizing a population's distribution of exposure duration, and may provide an estimate of intensity of the exposure. In addition, questionnaires offer the opportunity to collect data on multiple hazards, including those that are difficult to measure (e.g., musculoskeletal stressors) and those which can only be directly ascertained through individual experience (e.g., psychosocial stressors).

Some general idea of the prevalence of psychosocial stressors is available from existing surveys (NHIS, BRFSS, and QWS (NIOSH, 2013). NHIS and BRFSS have sufficient sample size to provide a reasonable level of detail on occupation but are fairly limited in detail on the stressors. The QWS is a special module assessing the quality of work life in America that has been added to the General Social Survey, a biannual, nationally representative, personal interview survey of U.S. households conducted by the National Opinion Research Center and funded by the National Science Foundation.

The module has been fielded every four years.[11] The QWS provides excellent detail on the stressors but the sample is small and does not provide sufficient detail about occupation. Experience with these, however, would need to inform approaches to survey of these stressors that could produce sufficient detail on both occupation and stressors to enhance surveillance of work and poor mental health.

Questionnaire surveys also offer respondents the opportunity to report on their "usual" exposure experience, while individual measurements observe only the condition at a specific time or day, and thus either miss less frequent conditions or require large sample sizes to capture them. Employee responses are valid when it comes to most of the questions on these surveys because they deal with attitudes or experiences.

An additional key advantage of questionnaire surveys is that they are relatively inexpensive to conduct. Furthermore, surveys on exposures can be added into other ongoing population-based surveys and/or conducted at regular intervals to provide information on changes in work conditions over time. Challenges in using questionnaires include reporting bias, employee awareness of exposures (e.g., low-level gaseous exposures may not be detected), and lack of quantitative information on the intensity of exposure. It may also be challenging to generalize across occupations.

Since 1991, Eurofound has been monitoring working conditions in Europe through its European Working Conditions Survey (Eurofound, 2017a,b). The survey aims to measure working conditions across European countries, identify groups at risk, and highlight concerns and progress, with the aim of contributing to European polices that would improve job quality (Eurofund, 2017b). In 2015, the sixth European Working Conditions Survey was conducted across 35 European countries and interviewed approximately 44,000 total employees and self-employed workers (Eurofund, 2017b). Workers were asked (in their native language) a range of questions concerning employment status, work organization, learning and training, working time duration and organization, physical and psychosocial risk factors, health and safety, work-life balance, worker participation, earnings and financial security, as well as work and health and trends in these exposures were reported.

## Next Steps

Several of the systems discussed above have potential for contributing to a comprehensive exposure and hazard surveillance. A system, properly structured, maintained, and funded, would fill important gaps in our ability to identify, monitor, and intervene on the myriad work-related risks,

[11] See http://gss.norc.org/Pages/quality-of-worklife.aspx.

especially those risks to long-term health of workers. Thus, the committee suggests that a comprehensive hazard and exposure system be developed and implemented, incorporating the strengths of the various approaches identified above, to collect, analyze, and distribute work-related risk information on a population basis. The creation of such a system will need to be undertaken with sensitivity to the need to protect the confidentiality of employees who participate in worksite monitoring. This issue has arisen in the past in settings where employees have expressed concern about risks to their employment relationship when negative exposure data are reported. This issue is one that employers, employees, and the agencies will need to bear in mind during the evolution of exposure surveillance activities.

The foundation of such a system would be a nationally representative survey of the working population in which many different types of work-related risks could be assessed for their prevalence, duration, and, crudely, for intensity. Only through such a system could workers in nonstandard employment arrangements be fully represented and avoid the challenges and limitations of employer-based surveys. In addition to classical dusts, chemicals, and physical and biological agents, emerging sources of risk such as psychosocial dimensions, shiftwork, and other organizational structures could be adequately described.

Such a population-based survey could be a stand-alone survey conducted by NIOSH, a part of the proposed HSOII survey, or as a periodic supplement to existing surveys such as the NHIS. It may be possible to collect a large sample at long intervals, e.g., every 10 years, or a smaller sample, appropriately stratified, at more frequent intervals, if the data can be integrated to provide an ongoing representative picture. Such a survey could be modeled on the European Working Conditions Survey, which has provided very substantial information for European policy makers and researchers.

Missing from a survey based on self-reported responses are quantitative exposure-level data, and risk factors that are not easily perceived by respondents. As discussed above, quantitative surveys of exposures at work are infeasible and limited for other reasons; however, ongoing exposure measurement activity by both government regulators and regulated industries provide the potential for development of exposure-level information, at least for selected highly regulated agents, and in some types of workplaces.

As described above, the compilation of MSHA particulate and noise-level information, and OSHA's integration of measurements into the IMIS database, has provided researchers the opportunity to extract and analyze exposure levels and trends while also investigating certain biases inherent in these compliance-based databases. Important limitations in these systems come from missing contextual information associated with the measurement data. In addition, employers collect a substantial amount of exposure-level information either independently or through use of consulting services.

These data rarely find themselves included in publicly available data resources due to privacy concerns. OSHA would need to explore ways these routinely collected data could be systematically collected and integrated with OSHA data while protecting the legal rights of the contributing workplaces. A precedent for this may be found in regulations concerning reporting injury and illness through the OSHA logs.

If these two systems—a periodic comprehensive population-based survey of working conditions and a publicly available compilation of exposure measurement data—could be linked together through common variables (e.g., industry, occupation, location, organizational type and size, etc.) the nation would begin developing a clear picture of exposures and, thus, work-related chronic disease risk and needs for disease prevention activities.

Each of these approaches offers opportunities for improving worker health through providing more detailed data on workplace hazards and hazardous exposures. Taking several concrete steps toward implementation would begin building a powerful system of risk identification and reduction. OSHA could enhance its IMIS by working with other regulatory agencies which already collect, or require collection of, exposure measurements including MSHA, DOE, state OSHA programs, and OSHA consultation programs. Federal research efforts (e.g., NIOSH health hazard evaluations) could also be included. In addition, research entities, especially those with federal funding, could be required to contribute their data to this system.

A major limitation to the utility of current IMIS data is missing data elements. The database design and data-collection tools would need to be updated to take advantage of modern informatics systems, in order to automate the entry of those elements which can be automated and reduce the data-entry requirements for compliance officers or inspectors. Use of a tablet-based system, integrated with laboratory and inspection-level data, could greatly enhance the comprehensiveness and completeness of the measurement database.

In the long term, a potential source of data would be from those workplaces required to collect exposure measurements in compliance with specific OSHA regulations. OSHA would need to consider amending its regulations to require such employers to contribute these data to the system. Provided to OSHA these data could be organized and analyzed to the benefit of the reporting employers. Feedback to individual employers could be provided that interpret the results in relative terms (benchmarking) and absolute terms (e.g., presence or frequency of exposures > Action Level).

Furthermore, OSHA could explore the development of a more comprehensive inspection observation tool, which could be used to describe the presence of multiple hazards in a workplace, and the distribution of exposure to these key agents. A predefined list of hazards could be developed, and inspectors could quickly estimate the number of employees potentially exposed, and the duration and frequency of exposure. This type of comprehensive data could be collected relatively easily through a tablet-based system. The burden on inspectors using such a tool would have to be addressed.

**Conclusion: The elements of a comprehensive exposure surveillance system can be largely achieved by building upon the exposure-related self-reported data envisioned in the expanded HSOII.** These data would provide population distributions of exposure to all types of hazards (chemical, physical, biological, ergonomic, physiological, psychosocial, and work organization), among workers in all types of working arrangements, and would include duration and frequency components of exposure. However, they would be limited in their ability to provide quantitative levels. The exposure intensity information could be integrated through the use of measurements, such as those collected in the OSHA IMIS. IMIS needs to be significantly enhanced to provide a more complete database of routinely collected measurement data.

**Recommendation H: NIOSH, in consultation with OSHA, should place priority on developing a comprehensive approach for exposure surveillance.** The objective should be to build systematically a comprehensive and continuously updated database of risks and exposures that provides the basis for estimating work-related acute and chronic health conditions for prevention.

In the near term:

- NIOSH should fully exploit the existing OSHA exposure databases by cleaning and integrating all available data sources to make them useful for surveillance purposes, taking proper account of the database limitations.

As an intermediate goal:

- NIOSH, in collaboration with OSHA and other agencies as appropriate, should construct an integrated exposure database to include the multiple sources of exposure measurement data already available, specifically MSHA's MSIS, Department of Energy and Nuclear Regulatory Commission personal exposure data, and relevant data from others conducting research with federal funds.

In the long term:

- NIOSH should link the integrated exposure database with the comprehensive survey data obtained in the recommended expanded HSOII

(Recommendation D) and new data from any characterization of exposures from targeted industry-specific assessments.

- NIOSH and OSHA should explore the feasibility of receiving employer-mandated exposure sample results after considering the reliability and quality of those measurements. The agencies should work with stakeholders to develop software and other tools and to facilitate establishment-level analysis of exposure data along with benchmarking.

## SUMMARY

This chapter has examined specific promising developments that, if strengthened and sustained, could significantly bolster occupational safety and health surveillance. The household survey has the potential to provide data about work from the perspective of the employees and thus supplement the current employer-based surveys. By including occupation and work-related information in electronic health records, meaningful links can be made between health and occupation with in-depth information on the health conditions. New methods for coding relevant to occupational health can expand the range of useful information by allowing the use of free-text documents as well as standardizing and assessing current efforts to extract and codify such information using emerging computational methods. Also pertinent are new employer-based electronic reporting initiatives that will provide greater data for tracking work-related illness and injury targeting and thereby enhancing prevention and treatment efforts. Worker's compensation data and programs contribute valuable occupational safety and health information, and enhanced use of those systems could help move surveillance efforts forward. Additionally, a number of public health surveys could provide valuable work-related information if questions or modules were added. Efforts by NIOSH to strengthen the leveraging of existing systems is key to cost-effective means of enhancing occupational safety and health surveillance. Development of a more comprehensive system for periodic collection and analysis of exposure-related information would greatly enhance the nation's ability to anticipate work-related health risks, especially for chronic diseases, which are less easily identified as being occupational in origin. Great strides could be made in this direction by improving the comprehensiveness and completeness of the exposure measurement databases already existing, and by incorporating exposure-related questions into a periodic population-based survey. Standardized, coordinated, and enhanced surveys, tools, and programs will allow occupational safety and health surveillance to become the fully functioning system that is needed to improve the health and well-being of workers. These promising developments will benefit from the research and new technologies discussed in Chapter 7.

## REFERENCES

AFL-CIO. 2016. *Death on the job: The toll of neglect. A national and state-by-state profile of worker safety and health in the United States, April.* Available online at https://aflcio.org/sites/default/files/2017-03/1647_DOTJ2016_0.pdf (accessed April 27, 2017).

Allen, A., L. Welch, K. Kirkland, D. Trout, and S. Baron. 2017. Development of a diabetes mellitus knowledge resource for clinical decision support assisting primary care physicians with work-related issues. *Journal of Occupational and Environmental Medicine* 59(11):e236-e239.

An, J. A., P. Braveman, M. Dekker, S. Egerter, and R. Grossman-Kahn. 2011. *Work, Workplaces, and Health.* Princeton, NJ: Robert Wood Johnson Foundation. Available online at http://www.rwjf.org/content/dam/farm/reports/issue_briefs/2011/rwjf70459 (accessed June 26, 2017).

Azaroff, L. S., C. Levenstein, and D. H. Wegman. 2002. Occupational injury and illness surveillance: Conceptual filters explain underreporting. *American Journal of Public Health* 92(9):1421-1429.

Baron, S., M. S. Filios, S. Marovich, D. Chase, and J. S. Ash. 2017. Recognition of the relationship between patients' work and health: A qualitative evaluation of the need for clinical decision support (CDS) for worker health in five primary care practices. *Journal of Occupational and Environmental Medicine* 59(11):e245-e250.

Bertke, S. J., A. R. Meyers, S. J. Wurzelbacher, A. Measure, M. P. Lampl, and D. Robins. 2016. Comparison of methods for autocoding causation of injury narratives. *Accident, Analysis, and Prevention* 88:117-123.

BLS (Bureau of Labor Statistics). 2013. *Support statement, survey of occupational injuries and illnesses, 2013.* Available online at https://www.reginfo.gov/public/do/PRAViewDocument?ref_nbr=201511-1220-004 (accessed January 30, 2017).

BLS. 2014a. *Occupational injury and illness classification manual.* Available online at https://www.bls.gov/iif/oshoiics.htm (accessed June 1, 2017).

BLS, 2014b. *Occupational Requirements Survey (ORS) consolidated FY 2014 feasibility tests summary report.* Available online at https://www.ssa.gov/disabilityresearch/documents/fy14_test%20summary.pdf (accessed November 8, 2017).

BLS. 2017a. *Standard Occupational Classification.* Available online at https://www.bls.gov/soc (accessed June 1, 2017).

BLS. 2017b. *Occupational Requirements Survey.* Available online at https://www.bls.gov/ors/home.htm (accessed November 8, 2017).

Bonauto, D., B. Silverstein, D. Adams, and M. Foley. 2006. Prioritizing industries for occupational injury and illness prevention and research, Washington State Workers' compensation claims, 1999-2003. *Journal of Occupational and Environmental Medicine* 48(8):840-851.

Bonauto, D. K., J. Z. Fan, T. W. Largo, K. D. Rosenman, M. K. Green, J. K. Walters, B. L. Materna, J. Flattery, T. St. Louis, L. Yu, S. Fang, L. K. Davis, D. J. Valiante, K. R. Cummings, J. J. Hellsten, and S. L. Prosperie. 2010. Proportion of workers who were work-injured and payment by workers' compensation systems—10 states, 2007. *Morbidity and Mortality Weekly Report* 59(29):897-900.

Boyer, J., M. Galizzi, M. Cifuentes, A. d'Errico, R. Gore, L. Punnett, and C. Slatin. 2009. Promoting Healthy Safe Employment (PHASE) in Healthcare Team. Ergonomic and socioeconomic risk factors for hospital workers' compensation injury claims. *American Journal of Industrial Medicine* 52(7):551-562.

CDC (Centers for Disease Control and Prevention). 2016a. Cardiovascular health status by occupational group—21 states, 2013. *Morbidity and Mortality Weekly Report* 65:794-798.

CDC. 2016b. Asthma among employed adults, by industry and occupation—21 states, 2013. *Morbidity and Mortality Weekly Report* 65:1325-1331.

CDC. 2017a. *About OIICS.* Available online at https://wwwn.cdc.gov/wisards/oiics/About.aspx (accessed June 1, 2017).

CDC. 2017b. What is PRAMS. Available online at https://www.cdc.gov/prams/index.htm (accessed June 12, 2017).

Cifuentes, M., J. Boyer, R. Gore, A. d'Errico, J. Tessler, P. Scollin, D. Lerner, D. Kriebel, L. Punnett, C. Slatin, and PHASE in Healthcare Research Team. 2007. Inter-method agreement between O*NET and survey measures of psychosocial exposure among healthcare industry employees. *American Journal of Industrial Medicine* 50(7):545-553.

Cifuentes, M., J. Boyer, D. A. Lombardi, and L. Punnett. 2010. Use of O*NET as a job exposure matrix: A literature review. *American Journal of Industrial Medicine* 53(9):898-914.

CSTE (Council of State and Territorial Epidemiologists). 2014a. 2013 National Assessment of Epidemiology Capacity: Findings and Recommendations. Available online at http://www.cste2.org/2013eca/CSTEEpidemiologyCapacityAssessment2014-final2.pdf (accessed November 8, 2017).

CSTE. 2014b. *Inclusion of work information as data elements in CDC surveillance systems.* Available online at http://c.ymcdn.com/sites/www.cste.org/resource/resmgr/2014PS/14_OH_02upd.pdf (accessed December 4, 2017).

CSTE. 2017. *Occupational health indicators.* Available online at http://www.cste.org/?OHIndicators (accessed June 8, 2017).

Das, R., J. McNary, K. Fitzsimmons, D. Dobraca, K. Cummings, J. Mohle-Boetani, C. Wheeler, A. McDowell, Y. Iossifova, R. Bailey, K. Kreiss, and B. Materna. 2012. Occupational coccidioidomycosis in California: Outbreak investigation, respirator recommendations, and surveillance findings. *Journal of Occupational and Environmental Medicine* 54(5):564-571.

d'Errico, A., L. Punnett, M. Cifuentes, J. Boyer, J. Tessler, R. Gore, P. Scollin, C. Slatin, and Promoting Healthy and Safe Employment in Healthcare Research Team. 2007. Hospital injury rates in relation to socioeconomic status and working conditions. *Occupational and Environmental Medicine* 64(5):325-333.

Eurofound (European Foundation for the Improvement of Living and Working Conditions). 2017a. *Sixth European Working Conditions Survey: 2015.* Available online at https://www.eurofound.europa.eu/surveys/european-working-conditions-surveys/sixth-european-working-conditions-survey-2015 (accessed June 1, 2017).

Eurofound. 2017b. *Sixth European Working Conditions Survey: Technical report.* Available online at https://www.eurofound.europa.eu/sites/default/files/ef_survey/field_ef_documents/6th_ewcs_-_technical_report.pdf (accesssed June 1, 2017).

Evanoff, B., A. Zeringue, A. Franzblau, and A. M. Dale. 2014. Using job-title-based physical exposures from O*NET in an epidemiological study of carpal tunnel syndrome. *Human Factors* 56(1):166-177.

Fan, Z. J., D. K. Bonauto, M. P. Foley, and B. A. Silverstein. 2006. Underreporting of work-related injury or illness to workers' compensation: Individual and industry factors. *Journal of Occupational and Environmental Medicine* 48(9):914-922.

Filios, M. S., E. Storey, S. Baron, G. B. Luensman, and R. N. Shiffman. 2017. Enhancing worker health through clinical decision support (CDS): An introduction to a compilation. *Journal of Occupational and Environmental Medicine* 59(11):e227-e230.

Frazier, T. 1983. NIOSH Occupational health and hazard surveillance systems. *Journal of Toxicology: Clinical Toxicology* 21(1-2):201-209.

Gardner, B. T., D. A. Lombardi, A. M. Dale, A. Franzblau, and B. A. Evanoff. 2010. Reliability of job-title based physical work exposures for the upper extremity: Comparison to self-reported and observed exposure estimates. *Occupational and Environmental Medicine* 67(8):538-547.

Harber, P., C. A. Redlich, S. Hines, M. S. Filios, and E. Storey. 2017. Recommendations for a clinical decision support system for work-related asthma in primary care settings. *Journal of Occupational and Environmental Medicine* 59(11):e231-e235.

HealthIT. 2017. *Meaningful use definition & objectives.* Available online at https://www.healthit.gov/providers-professionals/meaningful-use-definition-objectives (accessed June 12, 2017).

HHS (Department of Health and Human Services). 2015a. 2015 Edition Health Information Technology (Health IT) Certification Criteria, 2015 Edition Base Electronic Health Record (EHR) Definition, and ONC Health IT Certification Program Modifications, 80 *Federal Register* 16804-16921, March 30, 2015.

HHS. 2015b. "2015 Edition Health Information Technology (Health IT) Certification Criteria, 2015 Edition Base Electronic Health Record (EHR) Definition, and ONC Health IT Certification Program Modifications; Final Rule", 80 *Federal Register* 62602-62759, October 16, 2015.

Holloway-Beth, A., L. Forst, S. Freels, S. Brandt-Rauf, and L. Friedman. 2016. Occupational injury surveillance among law enforcement officers using workers' compensation data, Illinois 1980 to 2008. *Journal of Occupational and Environmental Medicine* 58(6):594-600.

HSE (Health and Safety Executive). 2017. *Data sources.* Available online at http://www.hse.gov.uk/statistics/sources.htm (accessed June 12, 2017).

Hutchings, S., J. Cherrie, M. van Tongeren, and L. Rushton. 2012. Intervening to reduce the future burden of occupational cancer in Britain: What could work. *Cancer Prevention Research* 5:1213-1222.

IAIABC (International Association of Industrial Accident Boards and Commissions). 2017. *EDI claims standards.* Available online at https://www.iaiabc.org/iaiabc/EDI_Claims.asp (accessed December 21, 2017).

IOM (Institute of Medicine). 2011. *Incorporating Occupational Information in Electronic Health Records: Letter Report.* Washington, DC: The National Academies Press.

Joe, L., R. Roisman, S. Beckman, M. Jones, J. Beckman, M. Frederick, and R. Harrison. 2014. Using multiple data sets for public health tracking of work-related injuries and illnesses in California. *American Journal of Industrial Medicine* 57(10):1110-1119.

Kauppinen, T., J. Toikkanen, D. Pedersen, R. Young, W. Ahrens, P. Boffetta, J. Hansen, H. Krumout, J. Maqueda Blasco, D. Mirabelli, V. de la Orden-Rivera, B. Pannett, N. Plato, A. Savela, R. Vincent, and M. Kogevinas. 2000. Occupational exposure to carcinogens in the European Union. *Occupational and Environmental Medicine* 57:10-18.

Kauppinen, T., S. Uuksulainen, A. Saalo, I. Mäkinen, and E. Pukkala. 2014. Use of the Finnish Information System on Occupational Exposure (FINJEM) in epidemiologic, surveillance, and other applications. *Annals of Occupational Hygiene* 58(3):380-396.

Kica, J., and K. D. Rosenman. 2014. Multi-source surveillance system for work-related skull fractures. *Journal of Safety Research* 51:49-56.

LA DH (Louisiana Department of Health). 2016. *Louisiana service worker wellness report, results from the Behavioral Risk Factor Surveillance System 2013 and 2014.* Available online at http://dhh.louisiana.gov/assets/oph/Center-PHI/BRFSS/data/LA_Service_Worker_Wellness_Report_BRFSS_2016_FINAL.pdf (accessed June 12, 2017).

LaMontagne, A.D., A. Milner L. Krnjacki, M. Schlichthorst, A. Kavanagh, K. Page, and J, Pirkis. 2016. Psychosocial job quality, mental health, and subjective wellbeing: A cross-sectional analysis of the baseline wave of the Australian Longitudinal Study on Male Health. *BMC Public Health* 16(Suppl 3):1049. doi: 10.1186/s12889-016-3701-x.

Largo, T. W., and K. D. Rosenman. 2015. Surveillance of work-related amputations in Michigan using multiple data sources: Results for 2006-2012. *Occupational and Environmental Medicine* 72:171-176.

Lavoue, J., M. Friesen, and I. Burstyn. 2013a. Workplace measurements by the U.S. Occupational Safety and Health Administration since 1979: Descriptive analysis and potential uses for exposure assessment. *Annals of Occupational Hygiene* 57(1):77-97.

Lavoue, J., M. Friesen, and I. Burstyn. 2013b. Workplace measurements by the U.S. Occupational Safety and Health Administration since 1979: Descriptive analysis and potential uses for exposure assessment. *Annals of Occupational Hygiene* 57(5):681-683.

Lupo, P. J., E. Symanski, P. H. Langlois, C. C. Lawson, S. Malik, S. M. Gilboa, L. J. Lee, A. J. Agopian, T. A. Desrosiers, M. A. Waters, P. A. Romitti, A. Correa, G. M. Shaw, L. E. Mitchell, and the National Birth Defects Prevention Study. 2012. Maternal occupational exposure to polycyclic aromatic hydrocarbons and congenital heart defects among offspring in the National Birth Defects Prevention Study. *Birth Defects Research Part A: Clinical and Molecular Teratology* 94:875-881.

MA DPH (Massachusetts Department of Public Health). 2016. *Putting data to work, 23 health indicators by occupation and industry: Findings from the Massachusetts Behavioral Risk Factor Surveillance System, 2012-2013*. Available online at http://www.mass.gov/eohhs/docs/dph/occupational-health/full-report-health-indicators-15.pdf (accessed June 12, 2017).

MacKenzie, W. R., A. J. Davidsony, A. Wiesenthal, J. P. Engel, K. Turner, L. Conn, S. J. Becker, S. Moffatt, S. L. Groseclose, J. Jellison, J. Stinn, N. Y. Garrett, L. Helmus, B. Harmon, C. L. Richards, J. R. Lumpkin, and M. F. Iademarco. 2016. The promise of electronic case reporting. *Public Health Reports* 131(6):742-746.

Marcum, J., and D. Adams. 2017. Work-related musculoskeletal disorder surveillance using the Washington State workers' compensation system: Recent declines and patterns by industry, 1999-2013. *American Journal of Industrial Medicine* 60(5):457-471.

Marucci-Wellman, H. R., T. K. Courtney, H. L. Corns, G. S. Sorock, B. S. Webster, R. Wasiak, Y. I. Noy, S. Matz, and T. B. Leamon. 2015. The direct cost burden of 13 years of disabling workplace injuries in the U.S. (1998-2010): Findings from the Liberty Mutual Workplace Safety Index. *Journal of Safety Research* 55:53-62.

Marucci-Wellman, H. R., H. L. Corns, and M. R. Lehto. 2017. Classifying injury narratives of large administrative data bases for surveillance. *Accident, Analysis, and Prevention* 98:359-371.

McLellan, R. K. , N. S. Haas, R. P. Kownacki, G. S. Pransky, J. B. Talmage, and M. Dreger. 2017. Using electronic health records and clinical decision support to provide return-to-work guidance for primary care practitioners for patients with low back pain. *Journal of Occupational and Environmental Medicine* 59(11):e240-e244.

Measure, A. C. 2014. *Automated coding of worker injury narratives*. Available online at https://www.bls.gov/iif/measure_autocoding.pdf (accessed June 1, 2017).

Meyer, J. D., M. Cifuentes, and N. Warren. 2011. Association of self-rated physical health and incident hypertension with O*NET factors: Validation using a representative national survey. *Journal of Occupational and Environmental Medicine* 53(2):139-145.

Monaco, K. 2016. Presentation to Committee on Developing a Smarter National Surveillance System for Occupational Safety and Health in the 21st Century, the National Academies of Sciences, Engineering, and Medicine, June 15, 2016, Washington, DC.

Nanda, G., K. M. Grattan, M. T. Chu, L. K. Davis, and M. R. Lehto. 2016. Bayesian decision support for coding occupational injury data. *Journal of Safety Research* 57:71-82.

National Commission on State Workmen's Compensation Laws. 1972. *Report of the National Commission on State Workmen's Compensation Laws*. Available online at http://workerscompresources.com/wp-content/uploads/2012/11/Introduction-Summary.pdf (accessed June 8, 2017).

NCCI (National Council on Compensation Insurance). 2017. *NCCI codes*. Available online at https://classcodes.com/ncci-codes (accessed June 1, 2017).

NCHS (National Center for Health Statistics). 2016. *NHIS supplements and co-sponsors*. https://www.cdc.gov/nchs/nhis/data-questionnaires-documentation.htm (accessed June 1, 2017).

NCVHS (National Committee on Vital and Health Statistics). 2012. *Development of Standards for the Collection of Socioeconomic Status in Health Surveys Conducted by the Department of Health and Human Services*. Letter to Kathleen Sebelius, Secretary, HHS, from Justine M. Carr, Chairperson, National Committee on Vital and Health Statistics, Washington, DC. June 22, 2012. Available online at https://www.ncvhs.hhs.gov/wp-content/uploads/2014/05/120622lt.pdf (accessed November 8, 2017).

NH HSS (New Hampshire Department of Health and Human Services). 2015. *Utilization of the NH Behavioral Risk Factor Surveillance System (BRFSS) to Better Understand Underreporting of Work-related Injuries*. Division of Public Health Services. Available online at https://iod.unh.edu/sites/default/files/media/NHOHSP/Pubs/utilizationnhbehavioralrisk.pdf (accessed June 12, 2017).

NM DH (New Mexico Department of Health). 2014. *Occupation and health in New Mexico*. Available online at https://nmhealth.org/data/view/report/162/ (accessed June 12, 2017).

NIOSH. 2012. *Guidelines for Reporting Occupation and Industry on Death Certificates*. DHHS (NIOSH) Publication No. 2012-149. Available online at https://www.cdc.gov/niosh/docs/2012-149/pdfs/2012-149.pdf (accessed April 27, 2017).

NIOSH. 2013. *Quality of worklife questionnaire*. Available online at https://www.cdc.gov/niosh/topics/stress/qwlquest.html (accessed April 17, 2017).

NIOSH. 2016a. *About NIOCCS*. Available online at https://www.cdc.gov/niosh/topics/coding/overview.html (accessed April 17, 2017).

NIOSH. 2016b. *How NIOCCS Works*. Available online at https://www.cdc.gov/niosh/topics/coding/how.html (accessed April 17, 2017).

NIOSH. 2016c. *NIOSH Center for Workers' Compensation Studies. May*. Available online at https://stacks.cdc.gov/view/cdc/39893 (accessed December 21, 2017).

NIOSH. 2017. Responses to questions from the National Academies committee about recommendations from the 1987 National Research Council report, January 24, 2017. Available through the National Academies Public Records Office.

NORC (National Opinion Research Center). 2016a. *Survey design and questionnaire for a Household Survey on Occupational Injuries and Illnesses: Report on suitability of existing surveys and frames (Deliverable #5)*. Available online at https://www.bls.gov/iif/norc-report-on-the-suitability-of-existing-surveys-and-frames.pdf (accessed June 1, 2017).

NORC. 2016b. *Survey design and questionnaire for a Household Survey on Occupational Injuries and Illnesses: Household survey design report (Deliverable #9)*. Available online at https://www.bls.gov/iif/norc-household-survey-design-report.pdf (accessed June 1, 2017).

NRC (National Research Council). 1987. *Counting Injuries and Illnesses in the Workplace: Proposals for a Better System*. Washington, DC: National Academy Press.

NRC and IOM (National Research Council and Institute of Medicine). 2001. *Musculoskeletal Disorders and the Workplace: Low Back and Upper Extremities*. Washington, DC: National Academy Press.

NY WCB (New York Workers' Compensation Board). 2017. *Description of manual classification code*. Available online at http://www.wcb.ny.gov/content/ebiz/eclaims/ReqTables/ManualClassificationCodes_DN0059.pdf (accessed June 1, 2017).

O'Halloran, A. C., P. J. Lu, W. W. Williams, P. Schumacher, A. Sussell, J. Birdsey, W. L. Boal, M. H. Sweeney, S. E. Luckhaupt, C. L. Black, and T. A. Santibanez. 2017. Influenza vaccination among workers—21 U.S. states, 2013. *American Journal of Infection Control* 45:410.

O*NET. 2017. *O*NET online*. Available online at https://www.onetonline.org (accessed June 12, 2017).

OSHA. 2016. Improve tracking of workplace injuries and illnesses. *Federal Register* 81(92): 29624-29693.

OSHA. 2017a. *Final rule issued to improve tracking of workplace injuries and illnesses*. Available online at https://www.osha.gov/recordkeeping/finalrule (accessed June 12, 2017).

OSHA. 2017b. *Establishments in the following industries with 20 to 249 employees must submit injury and illness summary (Form 300A) data to OSHA electronically*. Available online at https://www.osha.gov/recordkeeping/NAICScodesforelectronicsubmission.html. (accessed February 20, 2017).

Prang, K.-H., B. Hassani-Mahmooei, and A. Collie. 2016. Compensation Research Database: Population-based injury data for surveillance, linkage and mining. *BMC Research Notes* 9:456.

Rosenman, K. D., J. C. Gardiner, J. Wang, J. Biddle, A. Hogan, M. J. Reilly, K. Roberts, and E. Welch. 2000. Why most workers with occupational repetitive trauma do not file for workers' compensation. *Journal of Occupational and Environmental Medicine* 42(1):25-34.

Rushton, L., S. Hutchings, L. Fortunato, C. Young, G. S. Evans, T. Brown, R. Bevan, R. Slack, P. Holmes, S. Bagga, J. W. Cherrie, and M. van Tongeren. 2012. Occupational cancer burden in Great Britain. *British Journal of Cancer* 107:S3-S7.

Shockey, T. M., and A. G. Wheaton. 2017. Short sleep duration by occupation group—29 states, 2013-2014. *Morbidity and Mortality Weekly Report* 66:207-213.

Silverstein, B., E. Viikari-Juntura, and J. Kalat. 2002. Use of a prevention index to identify industries at high risk for work-related musculoskeletal disorders of the neck, back, and upper extremity in Washington State, 1990-1998. *American Journal of Industrial Medicine* 41(3):149-169.

Smith, C. K., and J. Williams. 2014. Work related injuries in Washington State's trucking industry, by industry sector and occupation. *Accident, Analysis, and Prevention* 65:63-71.

Spector, J. T., J. Krenz, E. Rauser, and D. K. Bonauto. 2014. Heat-related illness in Washington State agriculture and forestry sectors. *American Journal of Industrial Medicine* 57(8):881-895.

Towle, M., R. Tolliver, A. G. Bui, A. Warner, and M. Van Dyke. 2015. Adding industry and occupation questions to the Behavioral Risk Factor Surveillance System: New opportunities in public health surveillance. *Public Health Reports* 130:153-160.

TIRES. 2017. *TIRES prevents injuries in the trucking industry*. Available online at http://www.lni.wa.gov/Safety/Research/Trucking (accessed June 8, 2017).

UCLA (University of California Los Angeles). 2016. New on AskCHIS! Health, insurance data by industry and occupation. Center for Health Policy Research *News*, September 14. Available online at http://healthpolicy.ucla.edu/newsroom/press-releases/pages/details.aspx?NewsID=254 (accessed June 1, 2017).

U.S. Census Bureau. 2017a. *North American Industry Classification System*. Available online at https://www.census.gov/eos/www/naics (accessed June 1, 2017).

U.S. Census Bureau. 2017b. *Methodology*. Available online at https://www.census.gov/people/io/methodology (accessed June 1, 2017).

UT HD (Utah Department of Health). 2015. *Utah health status update: Industry and occupation impact on health*. Available online at http://health.utah.gov/opha/publications/hsu/1509_IndOcc.pdf (accessed June 12, 2017).

Utterback, D. F., and T. M. Schnorr, eds. 2010. *Use of Workers' Compensation Data for Occupational Injury and Illness Prevention: Proceedings from September 2009 Workshop*. Available online at https://www.cdc.gov/niosh/docs/2010-152/pdfs/2010-152.pdf (accessed June 8, 2017).

Utterback, D. F., and T. M. Schnorr, eds. 2013. *Use of Workers' Compensation Data for Occupational Safety and Health: Proceedings from June 2012 Workshop*, DHHS/CDC/NIOSH, Publication No. 2013-147. Available online at https://www.cdc.gov/NIOSH/docs/2013-147/pdfs/2013–147.pdf (accessed June 8, 2017).

van Tongeren, M., A. S. Jimenez, S. J. Hutchings, L. MacCalman, L. Rushton, and J. W. Cherrie. 2012. Occupational cancer in Britain: Exposure assessment methodology. *British Journal of Cancer* 107:S18-S26.

VT DH (Vermont Department of Health). 2016. *General worksite wellness—Data brief (4 of 4), 2014 Vermont Behavioral Risk Factor Surveillance System*. Available online at http://www.healthvermont.gov/sites/default/files/documents/2016/12/data_brief_201611_worksite_genl.pdf (accessed June 12, 2017).

WA DLI (Washington State Department of Labor and Industries). 2016. *BRFSS 2012-2014: Demographic and employment characteristics of cell phone only respondents*. Available online at http://www.lni.wa.gov/safety/research/files/96_1_2016brfss.pdf (accessed June 12, 2017).

WCIO (Workers Compensation Insurance Organizations). 2017. *Injury description table*. Available online at https://www.wcio.org/Document%20Library/InjuryDescriptionTablePage.aspx (accessed June 1, 2017).

Weiss, N. S., S. P. Cooper, C. Socias, R. A. Weiss, and V. W. Chen. 2015. Coding of Central Cancer Registry industry and occupation information: The Texas and Louisiana experiences. *Journal of Registry Management* 42(3):103-110.

WHO (World Health Organization). 2016. *Classifications*. Available online at http://www.who.int/classifications/icd/en (accessed June 1, 2017).

Wilkinson, R., and M. Marmot. 2003. *Social Determinants of Health: The Solid Facts*, 2nd ed. Copenhagen: WHO. Available online at http://www.euro.who.int/__data/assets/pdf_file/0005/98438/e81384.pdf (accessed June 19, 2017).

Wurzelbacher, S. J., I. S. Al-Tarawneh, A. R. Meyers, P. T. Bushnell, M. P. Lampl, D. C. Robins, C. Y. Tseng, C. Wei, S. J. Bertke, J. A. Raudabaugh, T. M. Haviland, and T. M. Schnorr. 2016. Development of methods for using workers' compensation data for surveillance and prevention of occupational injuries among state-insured private employers in Ohio. *American Journal of Industrial Medicine* 59(12):1087-1104.

Yamin, S. C., A. Bejan, D. L. Parker, M. Xi, and L. M. Brosseau. 2016. Analysis of workers' compensation claims data for machine-related injuries in metal fabrication businesses. *American Journal of Industrial Medicine* 59(8):656-664.

Zilaout, H., J. Vlaanderen, R. Houba, and H. Kromhout. 2017. 15 years of monitoring occupational exposure to respirable dust and quartz within the European industrial minerals sector. *International Journal of Hygiene and Environmental Health* 220(5):810-819.

# 7

# Key Actions to Move Forward with an Ideal National Occupational Safety and Health Surveillance System

Throughout the report, the committee uses a framework that describes a surveillance system as a set of processes that are enabled by different components. In the previous chapter, promising ongoing surveillance activities were reviewed and situated within this framework. This chapter examines four actions essential for moving forward with the national occupational safety and health (OSH) surveillance system that is needed to improve worker safety and health:

1. Set forth a clear rationale and prioritization for surveillance including quantifying the economic and health burden of occupational illness and injury;
2. Coordinate surveillance strategies and operations among key agencies, organizations, and stakeholders to ensure a "system-of-systems" architecture and align strategic planning and operations;
3. Use information technology effectively to meet surveillance objectives; and
4. Enhance training and support for surveillance practitioners with appropriate skills and knowledge to conduct, analyze, and recommend worker safety improvements based on OSH surveillance.

## SET FORTH A CLEAR RATIONALE AND PRIORITIZATION FOR OSH SURVEILLANCE

The committee's deliberations were guided by the premise that a 21st-century OSH surveillance system needs to collect, interpret, and analyze relevant data at the lowest cost feasible, and then rapidly and effectively disseminate that information to those who need to know. Meeting this goal requires a system that is *cost effective*; i.e., it will need to collect data that could have the biggest impact on improving worker safety and health, while minimizing the cost of collecting such data, and avoiding the collection of data that hold little value (Gold et al., 1996; Haddix et al., 1996). A cost-effective surveillance system balances the importance of collecting data about factors that can be modified to improve OSH with the costs of collecting each data element. It also measures costs and benefits from a *societal perspective,* consistent with the standards that have been widely accepted for cost-effectiveness analyses (Gold et al., 1996; Haddix et al., 1996; Muennig, 2008). This section examines four steps toward providing the basis and priorities for national OSH surveillance:

- quantifying the health and economic burdens,
- enhancing return from current resources,
- effectively allocating resources (evidence-based), and
- quantifying the fiscal rationale for coordination.

### Quantifying the Health and Economic Burden

As defined in Chapter 1, the *costs* of a surveillance system include not only the costs of conducting surveillance activities, but also the costs associated with the health and productivity consequences of occupational exposures, injuries, illnesses, and mortality on workers, their families, and society. Currently, there is no regular, standardized reporting on the overall economic burden of occupational illness, injury, and death in the United States, even though the most recent independent study estimated the burden to be $250 billion annually (Leigh, 2011). Given the enormity of this recent estimate, a regular national report on the financial and health burdens of occupational illnesses, injuries, and fatalities is essential for monitoring whether the United States is making measurable and meaningful progress in improving worker safety over time.

Resources to produce a national estimate on a regular basis could overcome some of the current methodological and data limitations. For example, the method for attributing mortality and morbidity to workplace illness and injury, known as attributable fractions, can lead to very different estimates of economic and health burden relative to official counts of workplace injuries, illnesses, and deaths, which are substantially lower than the estimates based on attributable fractions (Steenland et al., 2003).

The authors of a recently published, comprehensive framework for defining and measuring the health and economic burden of occupational "injury, disease, and distress" in the United States concluded that "the importance of documenting burden is to use it to plan investment in occupational safety and health risk prevention, risk management, and research and to prompt decision-makers to allocate funds for such investments" (Schulte et al., 2017). Likewise, the European Union (EU) has recently acknowledged the importance of regularly assessing the economic burden of occupational illnesses, injuries, and fatalities for Europe. The EU is currently implementing a process to routinely measure this burden and has recently published its first report on the availability and quality of data for developing European estimates of the burden of occupational illnesses, injuries, and fatalities (EU-OSHA, 2017).

An improved OSH surveillance system is the tool needed by policy makers, industry, workers, and other stakeholders to prioritize and target interventions that have been shown through research or practice to be most effective in improving worker safety and health. The surveillance system can then be used to monitor progress toward reducing the total economic burden of occupational illnesses, injuries, and fatalities over time.

**Conclusion: Identifying the areas of greatest need for OSH interventions through use of an improved national surveillance system and then targeting effective OSH interventions based on previous research and evidence is likely to produce significant and substantial savings to employers, employees, and society and increase worker health and well-being.**

If an improved national surveillance system is intended to inform actions to improve OSH outcomes, it is important to assess the value of both the investments in an improved surveillance system as well as the OSH outcomes. Cost-effectiveness analysis (CEA) is commonly used to measure the "value" of investments to improve health because it measures health outcomes in units other than dollars (Gold et al., 1996). The problem with using classical methods such as return on investment or cost-benefit analysis in health settings is that improvements in health must be valued in dollars. This requirement creates an ethical dilemma, because rather than value all lives equally, return on investment and cost-benefit calculations value lives differently, usually based on the wages or productivity of individuals in the economy. As a result, children and seniors are valued less, and high-income members of the workforce have the highest value. This limitation can be overcome by assigning average economic values to a life. For example, the Environmental Protection Agency uses mortality risk reduction, based on the value of a statistical life rather than the value of individual lives, when conducting analyses of societal costs related to exposure to environmental hazards (EPA, 2010). CEA, in contrast, places an equal value on health outcomes, regardless of the economic productivity of those who benefit, and thus avoids explicit valuation of life.

CEA generally measures health outcomes as lives saved, increased life expectancy (i.e., years of life saved), or reductions in adverse health outcomes (e.g., reduced incidence of injury). Cost-utility analysis (CUA) is a special case of CEA in which health outcomes are measured using quality-adjusted life years (QALYs) (Gold et al., 1996). QALYs capture both improvements in life expectancy as well as the quality of life for those years—that is, QALYs account for both mortality and morbidity (Gold et al., 1996). Social return on investment (SROI) has been proposed as an alternative method for more broadly valuing public health outcomes (Banke-Thomas et al., 2015), but the scientific evidence base for this method is not nearly as well-established as the broad evidence base for CEA and CUA.

**Conclusion: Because cost-benefit analysis and return on investment analysis value outcomes in dollar terms only, the committee finds that CEA and CUA are more appropriate methods for determining whether OSH surveillance systems produce value, i.e., improvement in health outcomes at a reasonable cost.** CEA and CUA are preferred because they give equal weight to all workers, in contrast to other methods, which value individuals based on their salaries or wages.

An assessment of OSH surveillance systems, therefore, should measure and balance the costs of data collection, analysis, and dissemination versus measurable reductions in mortality and morbidity (including time lost from work) that can be attributed to actions taken based on OSH surveillance.

### Enhancing Return from Current Resources

Regular national reporting on the economic burden of occupational illnesses, injuries, and fatalities can also assist federal agencies, state and local governments, employers, and employee groups to address surveillance objectives more effectively with existing resources. Burden-of-illness studies have been conducted in the United States for decades and have estimated the economic burden of diseases for the entire population (e.g., Rice, 1967; Cooper and Rice, 1976; Rice et al., 1985); for specific conditions (e.g., Thorpe et al., 2015); or for certain outcomes, such as fatalities (Luo and Florence, 2017). In recent years, federal agencies such as the National Institutes of Health have called for greater use of such studies in prioritizing areas for funding to combat heart disease, cancers, diabetes, etc., although empirical studies suggest at most only a moderate association between funding priorities and population disease burden (Gross et al., 1996; Gillum et al., 2011). Existing federal efforts such as the Centers for Disease Control and Prevention's (CDC's) Web-based Injury Statistics Query and Reporting System (WISQARS) provide another avenue for enhancing existing resources. WISQARS provides an interactive, online database of fatal and nonfatal injury, violent death, and cost-of-injury data that could be modified to report separately on occupational illnesses, injuries, and fatalities.

Finally, a regular report on the economic burden of occupational illnesses, injuries, and fatalities could serve an important role in guiding national funding priorities to ensure that current funding is being targeted cost effectively, i.e., where the burden is greatest, where effective interventions exist to reduce work-related mortality or morbidity, and where intervention costs, including the costs of regulation, are minimized in producing improved occupational health outcomes.

### Effective (Evidence-Based) Allocation of Resources

Another benefit of a regular national report on the economic burden of occupational illnesses, injuries, and fatalities would be to establish funding priorities by federal agencies with OSH responsibilities. The challenge when new resources become available or resources are newly constrained is to determine the most cost-effective way to allocate changing resource levels. Agencies can use expert panels and the evidence base established from previous studies to determine where effective interventions exist for reducing adverse outcomes, and to target those conditions where health outcomes can be improved in the most cost-effective manner.

### Quantification of Fiscal Rationale for Coordination

Because there is no single government agency responsible for all aspects of OSH, coordination across agencies is essential to avoid unnecessary duplication of effort and to maximize efficient and effective use of existing resources. The Occupational Safety and Health Administration (OSHA) has primary responsibility for federal OSH activities, but the National Institute for Occupational Safety and Health (NIOSH), the Bureau of Labor Statistics (BLS), other federal agencies, and state agencies play important roles as well through surveillance and the surveys they administer. The Department of Health and Human Services (HHS) maintains national health accounts that collect data to monitor trends in national spending for health care services. Some portion of the spending is attributable to occupational illnesses, injuries, and fatalities, but HHS national health accounts do not separately identify OSH costs and expenditures. Better coordination could lead to efforts to partition national health spending into occupational and non-occupational health spending estimates.

**Recommendation I: NIOSH should coordinate with OSHA, BLS, and other relevant agencies to measure and report, on a regular basis, the economic and health burdens of occupational injury and disease at the national level.** This report should also attempt to address the contribution of implemented interventions in reducing these burdens. The advantages of a regular, standard report on national economic burden of occupational injury and disease include:

- focusing attention on the significant burden that already exists,
- measuring progress over time in reducing those burdens and improving worker safety and health,
- improving the allocation of existing resources to improve health outcomes, and
- establishing priorities.

Research, such as to establish the fraction of disease and injuries attributable to occupational exposures, will be necessary to continually improve the quality of burden estimates that can be produced.

## COORDINATE SURVEILLANCE STRATEGIES AND OPERATIONS AMONG KEY AGENCIES, ORGANIZATIONS, AND STAKEHOLDERS

As was described in Chapter 1, surveillance entails the collection and analysis of data, followed by the interpretation and dissemination of information to relevant actors to meet public health and prevention objectives. The legal and organizational context in the United States, and to a lesser extent in other countries, is such that many aspects of OSH surveillance are under the mandate of different agencies, which may have different objectives. However, to have an efficient OSH surveillance system for the country, surveillance activities must be coordinated within and among the different agencies to allow the attainment of national objectives, while respecting and further advancing the objectives of each agency. There are three ways in which improvements could be made:

- Implement a system of systems approach,
- Improve the alignment of existing systems, and
- Coordinate communications.

### Implement a System of Systems Approach

One approach to coordinating OSH surveillance at the national scale is to create a "system of systems," which has been defined as "a collection of task-oriented or dedicated systems that pool their resources and capabilities to obtain a new, more complex 'meta-system' which offers more functionality and performance than simply the sum of the constituent systems" (Popper et al., 2004). Such a system is created by connecting otherwise independent systems, which have operational and managerial independence, are geographically distributed, and are heterogeneous (Maier, 1998). In contrast to large, monolithic systems that are controlled centrally through a hierarchical structure with clear lines of reporting (Fisher, 2006), systems of systems are characterized by distributed control and cooperation (Keating et al., 2003). In response to the 2009 influenza pandemic, this strategy was used to rapidly develop a national syndromic surveillance system covering more than half of all emergency department visits in the country (Olson et al., 2011). In that example, city and state syndromic surveillance systems adopted a common model for reporting aggregated data to a central system, which then combined the data to produce summaries and identify trends at a national scale.

**Conclusion: Given the current context of OSH surveillance, development of a national "system of systems" is more likely to result in rapid and sustainable gains in OSH surveillance as compared to development of a new national monolithic system, which is likely to require substantial legislative changes and substantial additional resources.**

### Improve the Alignment of Existing Systems

Although a "system of systems" is a compelling strategy for developing a national OSH surveillance system, creating such a system can be complicated by technical, human, and organizational differences among existing systems (Wells and Sage, 2008) as well as regulatory impediments. These complications are likely to be more pronounced if the intent is to couple individual systems tightly. However, a loose coupling is possible, where the architecture of the overall system is essentially a set of standards that allow meaningful communication or exchange of data and information among systems (Maier, 1998). For example, the influenza syndromic surveillance system described earlier used a loose coupling, requiring agreement only on a form for reporting aggregated data. The distinction between data[1] and information[2] is important, because data, especially at the individual level, are often difficult to share across organizations, while information is generally easier to share. Ultimately, what must drive the design of a system of systems, including which data and information should be shared, are the OSH objectives that are only possible or feasible to attain through interoperation of otherwise independent systems. The new information that the national system of systems could produce would be an "emergent property," or a novel contribution to OSH surveillance not feasibly available from

---

[1]Data are raw facts, which generally afford little insight on their own (e.g., the age of a worker).

[2]Information is obtained by placing data in context, for example through analysis or combination with other data (e.g., counts or rates of injuries in a specific age group).

any existing system on its own but attainable through a national system. This new information should allow objectives described in Chapter 2 to be achieved more completely and more efficiently. For example, enhanced data and information sharing through coordinated planning and operations of existing surveillance systems should allow more accurate measurement of the burden of work-related illness, enable clearer identification of working populations at high risk for work-related injury or illness, and provide richer data to generate hypotheses and conduct epidemiological research.

### Coordinate Communication of Information from Surveillance

An essential part of surveillance is ensuring that the information, produced by analyzing and interpreting collected data, is disseminated to stakeholders who are in positions to use that information to protect and improve worker safety and health. As described in Chapter 3, these stakeholders are many and include policy makers; federal, state, and local government agencies; individual employers and workers; industry and worker organizations; insurers; health and safety professionals; educators; and researchers, as well as the health care community. Timeliness in using surveillance results for planning and action, a key component in the evaluation of a surveillance system (CDC, 2001), means providing outputs from surveillance in usable formats that meet the needs of the information user(s).

Ongoing dissemination of surveillance information takes many forms, ranging from annual reporting of aggregated data to placing surveillance indicators on interactive websites and making them available through user queries. BLS, NIOSH, and OSHA currently have separate websites where they post surveillance findings (BLS, 2017a,b; NIOSH, 2017a,b,c,d,e,f,g,h; OSHA, 2017). Each of these sites has interactive components. BLS issues press releases and provides annual reports for the Census of Fatal Occupation Injuries (CFOI) and the Survey of Occupational Injuries and Illnesses (SOII). More extensive analyses of surveillance data are published in BLS periodicals. New findings from NIOSH and other agencies are periodically posted but with irregular time periods between postings. NIOSH findings based on more extensive analysis of surveillance data are published in peer-reviewed journals. The NIOSH surveillance home page includes links to NIOSH webpages for its various surveillance systems and functions that enable the user to query BLS employment, CFOI and SOII data, and several of the NIOSH surveillance data sources.

NIOSH also maintains a clearinghouse for surveillance reports and educational materials and surveillance tools generated by all the state-based surveillance programs. While these web-based resources represent a significant improvement in information dissemination in recent years, there is currently no central site or compilation of the information produced by the various agencies. Even within the same agency there are separate websites for different conditions (e.g., NIOSH: respiratory disease, and state-based surveillance; NIOSH, 2017a,b,c,d,e,f,g,h). This contrasts with the dissemination of data in Great Britain by their Health and Safety Executive, where, for example, both employer and worker survey data are available through a single site (HSE, 2016).

As with any surveillance system, a national OSH system of systems must also communicate information to stakeholders, ideally in a manner that is consistent with the reporting of information by the OSH agencies that operate component surveillance systems. Therefore, in addition to coordinating the establishment of objectives and the technical activities for a national OSH surveillance system, communication from the national system should also be coordinated in an ongoing manner.

The information generated by surveillance systems includes measures, such as rates of exposures and outcomes. To be clear, centralizing information does not imply centralizing data, which is impractical and unnecessary on a national scale. Information can be centralized, for example, by having each participating system make the information it intends to disseminate available in a standard, machine-readable format, which would allow a central system to access and integrate information from different systems routinely in an automated manner. Once the information is in one location, it can be made available through a clearinghouse such as the one established by NIOSH. Once information is centralized, it is also possible to coordinate more urgent dissemination activities, such as through the health alert network maintained by the CDC. The committee returns to the topic of dissemination of surveillance information in the following section, where we consider the potential for informatics to enable OSH surveillance processes, including dissemination.

## USE INFORMATION TECHNOLOGY EFFECTIVELY

An informatics perspective, where the focus is on the optimal use of data and knowledge to meet OSH surveillance objectives, allows a principled assessment of the potential benefit of new technologies to enable OSH surveillance processes. The ongoing and unprecedented gains in fundamental technologies for storing, communicating, and analyzing data have produced a dizzying array of methods and tools with the potential to automate, advance, or replace many existing processes in OSH surveillance, including data collection, data management, analysis, and dissemination. One strategy for making sense of these many opportunities is by viewing OSH surveillance through the lens of biomedical informatics, asking how data and knowledge can be used optimally to solve problems and make decisions in OSH surveillance (Shortliffe, 2014). From this perspective, the

OSH surveillance processes and objectives are central and it is possible to consider, ideally from an evidence-based perspective, which innovations in information technology are likely to make surveillance processes (data collection, analysis and interpretation, and dissemination of information) more efficient and effective. There is limited direct evidence in the biomedical literature about the effective use of existing and emerging information technologies for OSH surveillance, but it is possible to draw relevant insights from the literature on public health surveillance, from biomedical informatics more generally, and from advances in data management and analysis in domains beyond health and health care (e.g., electronic health records, machine learning, and social networks).These insights are important in identifying potential roles of new technologies in individual OSH surveillance systems and to understand how these technologies can contribute to the development of a national OSH surveillance system.

### Building Capacity in Information Technology Expertise

Even when there is evidence that new information technologies can be beneficial, it can be challenging to achieve these benefits in practice. Expertise in information technology (IT) and informatics must exist within OSH surveillance agencies to realize the potential benefits. People with these skills and knowledge are critical for reviewing technical evidence, focusing attention on OSH priorities driving investments in information technology, and ensuring successful implementation of adopted technologies. Even if agencies make extensive use of contract agencies to deliver IT services, it remains essential to have a core complement of biomedical informatics expertise to set strategy and to ensure the effective management of contracted resources. Several academic institutions offer formal graduate degrees in biomedical informatics, emphasizing a range of competencies that are well matched with the needs of the OSH community (Kulikowski et al., 2012).

For a variety of reasons, however, it is difficult for OSH agencies to hire and retain staff with the knowledge and skills needed to plan, manage, evaluate, and use new information technologies. Consequently, agencies tend to employ fewer staff members with informatics expertise than they need and the turnover among these employees can be high. Turnover is problematic because most people with informatics expertise will know little about OSH when they join an agency, but over time they will acquire deeper knowledge and experience in OSH, thereby becoming increasingly valuable to the agency.

There is an acknowledged shortage of expertise in biomedical informatics, especially as it relates to public health surveillance (Edmunds et al., 2014). In this context, OSH agencies must be strategic in how they recruit, cultivate, and retain this expertise. In the longer term, supporting expansion of training programs may be beneficial. In the medium term, increasing use of internship and fellowship opportunities, such as the public health informatics fellowships at CDC, can introduce people with biomedical expertise to the challenges and opportunities in OSH surveillance.

**Recommendation J: NIOSH should build and maintain a robust internal capacity in biomedical informatics applied to OSH surveillance.**

In the near term:

- Assess the need within the agency for expertise in biomedical informatics in the context of current and future demand, recognizing that it will be important to train informatics talent in OSH surveillance and then to work to retain talented individuals who develop knowledge at the intersection of the informatics discipline and OSH applications;
- Create an organizational strategy for deploying and making optimal use of expertise in biomedical informatics to support the planning and conduct of OSH surveillance;
- Develop a plan for hiring, including consideration of steps such as reaching out to academic programs, advertising in different venues, and offering internships; and
- Develop a plan for retention, including opportunities for continuing education.

## Systems Architecture and Overall System Functioning

Next steps forward for OSH surveillance need to focus on technologies that can contribute to the overall functioning of a surveillance system by supporting activities across multiple processes. For example, standard controlled terminologies enable the consistent representation of data within many surveillance processes and the communication of data between processes within a system. Technologies for integrating data across systems, or allowing distributed analysis across systems, is another important technology within a national system—something that would not be possible with any single system.

### *Controlled Terminology Development and Standards and Harmonization*

Controlled terminologies (or, data standards) can ensure that a given concept is recorded consistently by different people over time within and across systems by enumerating the accepted ways for concepts to be encoded. These terminologies enable communication between systems, with interoperability achieved more easily when different systems use the same controlled terminologies. Interoperability between systems, however, also requires agreement on a communication or messaging standard. In other words, a controlled terminology defines what is in a message, and the messaging standard defines the structure of the message.

As reviewed in Chapter 6, many organizations have developed or adopted a number of OSH controlled terminologies in support for their activities. For the most part, these terminologies have evolved independently to meet the needs of different organizations. While each terminology represents some aspect of OSH, when taken together, the existing terminologies do not cover all of OSH in a consistent manner. These terminologies do however represent many concepts central to OSH surveillance, including the occupation of a worker (Standard Occupational Classification, or SOC) and the industry of a business establishment (the North American Industry Classification System, or NAICS). A smaller terminology was also created for use with the census by merging subsets of the SOC and the NAICS (Bureau of the Census codes). Controlled terminologies also are used to represent outcomes or events (i.e., injuries and illnesses) within OSH surveillance systems, including the International Classification of Disease (ICD, with modifier codes to represent details of an injury), the Occupational Injury and Illness Classification System (OIICS), and the Workers Compensation Insurance Organizations (WCIO) codes.

The Bureau of the Census codes are directly related to the SOC and NAICS codes, as they were explicitly derived from these other two coding systems. There are, however, no other existing mappings or explicit linkages between the various controlled terminologies used for OSH surveillance. Consequently, it is possible to encode combinations of occupation and industry which are not possible in practice. Another consequence is that outcomes coded using different controlled terminologies are not easily compared (e.g., ICD and OIICS, WCIO and OIICS), hindering the integration of data across different surveillance systems in the absence of crosswalks between coding systems (Koeman et al., 2013).

One possible solution to this problem is to develop and maintain a mapping between these related terminologies, or what is called a metathesaurus (Harber and Leroy, 2017). An example in the biomedical domain is the Unified Medical Language System (UMLS), a compendium of controlled terminologies developed and maintained by the National Library of Medicine of the National Institutes of Health, which links related concepts or terms across dozens of controlled terminologies. Of the controlled terminologies discussed above, only the ICD is included in the UMLS.

**Recommendation K: NIOSH should work with the National Library of Medicine to incorporate core OSH surveillance terminologies, including those for industry and occupation, into the Unified Medical Language**

**System (UMLS).** The creation and maintenance of mappings among OSH terminologies and between OSH terminologies and other relevant terminologies already included in the UMLS should be considered.

In the near term*:*

- Establish an inventory of relevant OSH terminologies;
- Develop use cases that benefit from the existence of mappings across OSH terminologies; and
- Prioritize terminologies in terms of the value that accrues from incorporating them into the UMLS.

In the longer term:

- Incorporate highest-priority OSH terminologies into the UMLS.

One topic for which data representation standards are not currently available is public health interventions. While terminologies do exist for recording clinical interventions that target individual patients (Hanser et al., 2009), representation standards for interventions that target workplaces or populations are not available. A standard for these interventions would allow public health officials and others to record actions taken to prevent and control occupational injury and disease in a systematic manner. While there have been calls for more consistent reporting of interventions in epidemiological studies (Des Jarlais et al., 2004), this approach has not been adopted widely by public health agencies. Efforts are now under way to develop machine-readable representations of interventions to support the consistent recording of behavioral (Michie et al., 2013) and public health interventions (Shaban-Nejad et al., 2017). Adoption and use of these ontologies to record when and where interventions are implemented should allow systematic monitoring of the effectiveness of occupational health interventions used in real-world settings. The resulting information could then be used to evaluate and continually refine these interventions in what would be a learning OSH system.

### *Data Integration and Storage*

An important part of a system of systems is to enable access to individual, case-level data in a manner that protects confidentiality but allows the identification of new hazards, and the introduction of old hazards into new industries. Once data are collected or obtained from another source for use within a surveillance system, the data must be stored and integrated, or linked. Within a single system, the data are usually stored in one location, often within a database or a data warehouse, where the data have been arranged to optimize regular queries. The most informative linkages between data sources are individual-level linkages (e.g., linking data on an individual's occupation and industry with data on their health outcomes and time away from work). Linkages based on location (e.g., linking home address to census variables) or other attributes (e.g., linking occupation to an exposure matrix) may also be useful for data analysis.

Data integration and storage can pose challenges within a single surveillance system, and these challenges are greater for a system of systems. In a national system, it is not feasible to store all data in a single location due to legal barriers and agency policies that limit data sharing. However, a sufficient degree of data integration may still be possible without centralizing data. With a federated data strategy, a "virtual" database can be created by identifying the linkages between databases at each participating location. Queries made against this virtual database are answered by accessing data from the relevant locations and assembling them into a single response. It is possible with this approach for each data provider to control which data are visible to the larger system and to approve or deny any query.

Another related strategy is to perform distributed analyses, as opposed to distributed queries. With this approach, the data remain with the individual systems, and statistical analyses are distributed across participating agencies (Gini et al., 2016). For example, cohorts of workers in an industry could be identified across multiple surveillance systems and the overall effect of an exposure in that industry could be estimated by pooling the results of the same statistical analysis performed by each system against its own data.

In summary, multiple strategies exist for deriving additional value by querying and analyzing data held by different OSH agencies. The preferred strategy must be developed to realize the objectives of a system of systems, while respecting limitations around data sharing and available resources. The topic of data integration and distributed analysis is discussed later in the chapter in the section considering how informatics can be used to support data analysis.

## Data Collection and Processing

The collection and processing of data in a surveillance system can benefit greatly from innovations in information technologies. Novel technologies such as environmental and personal sensors can be used to capture data. People can be empowered to collect data through crowdsourcing and data can be captured from posts to social media. Another rich source of data is the electronic health record (EHR), and novel methods for converting free text to structured codes can play an important role in processing EHR data to make them usable for OSH surveillance.

### *Mobile Devices and Sensors*

Mobile devices are now ubiquitous with adoption at 96 percent globally in 2014 and 65 percent of U.S. adults owning a smartphone in 2015 (Pew Research Center, 2015). These devices contain sensors, which can capture location, sound, images, acceleration, and other measurements. Moreover, the devices can then transmit those data wirelessly to a central data repository. In recognition of the potential impact of such distributed sensors, NIOSH developed the Center for Direct Reading and Sensor Technologies to work with partners in advancing the development and use of sensors for OSH (NIOSH, 2016).

One application of mobile devices is to gather data on occupational exposures. Data can be captured actively by having inspectors or employers use applications that capture and process data from the device sensors, then submit results. This approach is already used for medical or exposure monitoring, but these applications generally rely on specialized devices (Evans et al., 2010), especially in settings where there is the potential for remote devices to pose a hazard due to flammability or explosion. Increasingly sophisticated monitoring and even analysis is possible using mobile devices. For example, in the context of infectious disease surveillance, routine tests such as blood smears to detect parasites can be performed with acceptable accuracy using mobile devices (Pirnstill and Coté, 2015). From a chronic disease perspective, even simple mobile phones allow tracking of healthy behaviors and environmental exposures (Donaire-Gonzalez et al., 2016).

### *Crowdsourcing*

Crowdsourcing is a passive approach to data collection that empowers people to use apps on mobile devices to capture and then voluntarily forward data to a central site for analysis and dissemination (Brabham et al., 2014). Applications to enable crowdsourcing can be made generally available, for example to measure noise or air pollution, or they can be used by an employer for "internal crowdsourcing" (Brauch, 2015). People can be motivated to submit data using different strategies, including their engagement in what has been called participatory epidemiology (Freifeld et al., 2010) or citizen science (Pocock et al., 2017), where those who submit data are also engaged in their analysis and interpretation. Although this democratization of data access and analysis presents many opportunities, it also raises new questions such as how to derive unbiased insights from data collected through crowdsourcing (Welvaert and Caley, 2016) and how data informally collected through crowdsourcing or other means are best used by employers, employees, communities, and researchers in the absence of any authoritative interpretation.

### *Social Media*

Social media provide a platform for people to communicate their thoughts and experiences with members of their social network and others who may be interested. Soon after the widespread adoption of social media, researchers recognized the value of systematically monitoring these public communications to generate public health intelligence. An early application of this approach to surveillance was to monitor the incidence of influenza-like illness (Chew and Eysenbach, 2010) and it is clear now that social media data can improve surveillance of seasonal influenza epidemics when combined with more traditional surveillance approaches (Mitchell and Ross, 2016). More recently, this approach has also been used to monitor the frequency of adverse effects of prescription medications. Comparisons to traditional approaches, such as spontaneous reporting, suggest that social media surveillance can identify adverse events and their frequency of occurrence (Powell et al., 2016).

Influenza-like illness and adverse drug reactions are highly prevalent events, making them ideal outcomes for surveillance through social media. Individual types of occupational injury and illness tend to occur less frequently, so additional research is required to determine the extent to which these events can be monitored through social media.

### *EHRs and Electronic Reporting*

EHRs are increasingly used routinely in primary care and other settings, such as in emergency departments. These systems capture a range of data, which researchers have shown to be valuable for a variety of public health surveillance objectives. Syndromic surveillance is one example, where anonymized data on the reason for the encounter are collected for all patients, usually from emergency departments, and then analyzed to detect unusual increases in health care utilization for broad categories of symptoms, such as influenza-like illness and gastrointestinal disease (Mandl et al., 2004). More recently, these methods have been extended to allow automated, case-based surveillance, where more complex case definitions are applied to the multiple types of data integrated within an EHR, allowing the accurate and timely detection and reporting of cases of communicable diseases (Vogel et al., 2014).

As discussed in Chapter 6, there is an ongoing effort to increase the amount of occupational data recorded in the EHR and the growing adoption of data standards facilitates the analysis of these occupational data together with other types of data contained in the EHR. In particular, the electronic case-reporting initiative, to which NIOSH has contributed, has defined and begun to develop an infrastructure that will enable the automated reporting of health conditions by occupation (Mac Kenzie et al., 2016). However, the collection and reporting of occupation, industry, and other OSH data in EHRs remains voluntary. The quality of the data in EHRs can be variable, so methods are needed to assess and assure the quality of OSH data extracted from EHRs.

**Recommendation L: NIOSH should lead efforts to establish data standards and software tools for coding and using occupational data in electronic health records.** These efforts should be coordinated with the Office of the National Coordinator for Health Information Technology (ONC) to support the establishment of a rule requiring collection and effective use of OSH data in the electronic health record.

In the near term:

- Develop a consensus within the OSH surveillance community regarding the preferred terminologies and tools for extracting data on industry and occupation from the EHR;
- Engage with ONC to communicate this consensus to other stakeholders and to establish a broader consensus among all stakeholders regarding an acceptable strategy; and
- Support ONC in the process of establishing a rule to require the capture of industry and occupation in the EHR.

In the longer term:

- Work with the occupational medicine and general medicine community to develop models and tools for using occupational data in electronic health records for clinical care and for serving the prevention needs of the clinical population.

*Autocoding Software*

Natural language processing (NLP) is the field of computer science concerned with the interpretation by computers of natural language, including the automated interpretation of written text. When working with written text, NLP software attempts to link words in the written text to terms in a controlled vocabulary. In other words, the NLP programs are trying to automatically assign codes to the written text. Accordingly, the term "autocoding" is often used in the OSH surveillance literature to refer to the automated assignment of codes (e.g., occupation, industry, and type of injury) to words in a textual report using NLP software.

Historically, NLP software has relied on manually developed rules to map or link written words to standard terms. More recently, statistical methods have been used in NLP programs. The statistical methods for NLP learn a model from a set of written documents, which are already coded, and the statistical model is then used to predict the best coding for new documents. Both NIOSH and BLS have developed NLP software to support OSH surveillance. NIOSH has employed multiple strategies for autocoding, including developing and making available in 2012 a web-based NLP system (the NIOSH Industry and Occupation Computerized Coding System) that uses rules to assign industry and occupation codes (i.e., Bureau of Census codes with links to SOC and NAICS codes) to text in vital statistics, health survey, and electronic health records. Version 3 of this software is scheduled for release in 2018. In 2012, BLS began exploring the use of NLP and it developed a statistical NLP system. This system was used to assign SOC codes to text in responses to the 2014 SOII and was extended in 2015 to automatically code nature of injury and part of body affected in responses to the SOII (BLS, 2015). The Bureau of the Census has also developed autocoding strategies for employment data collected in the Current Population Survey and American Community Survey.

In addition to efforts by OSH agencies, academic groups (Patel et al., 2012; Burstyn et al., 2014; Harber and Leroy, 2017) have developed and evaluated NLP programs to code occupational history in free-text documents. Some general observations across all these efforts are that NLP programs are usually unable to code some records (30 to 50 percent) and for those that are coded, agreement with manual coding is reasonable (50 to 80 percent). Another observation is that a wide range of NLP programs have been developed, using different overall frameworks, algorithms, and lexicons. Although each approach has advantages, no single approach appears to be ideal for all types of documents in all settings and NLP is a fast-moving area of research.

**Conclusion: Using natural language processing to extract data from free text has the potential to improve the efficiency of surveillance in many ways.** There is also the potential to influence future rules if "industry standard" approaches can be identified for extracting OSH text from electronic health records.

**Recommendation M: NIOSH and BLS, working with other relevant agencies, academic centers, and other stakeholders should coordinate and consolidate, where possible, efforts to develop and evaluate state-of-the-art computational and analytical tools for processing free-text data found in OSH surveillance records of all types.** This coordination should enable rapid innovation and implementation, into OSH surveillance practice, of successful "autocoding" methods for different data sources.

In the near term:

- Conduct an inventory of activities and key stakeholders and
- Support knowledge exchange activities (symposia, competitions).

In the longer term:

- Develop open data sets that can be used to consistently evaluate methods for extracting OSH data from free text.

## Analysis and Interpretation

Data collected through surveillance needs to be analyzed and interpreted appropriately to generate information that surveillance stakeholders can use to guide their actions. In practice, a range of analytical strategies are applied to surveillance data, reflecting different objectives, data, and expertise. In many surveillance settings, descriptive analyses are performed periodically to identify trends over time or unexpected patterns in population subgroups. For example, NIOSH detected an increasing trend in early coal workers' pneumoconiosis, leading to a more detailed analysis at the state level, and ultimately preventive actions (see Box 7-1).

Periodic analysis of trends remains important, but advances in computational and statistical methods now make it possible to analyze large volumes of data and to detect meaningful variations in health outcomes that may warrant public health intervention (Lombardo and Buckeridge, 2007; Fricker, 2013). Although not widely applied in surveillance practice, researchers are also exploring the potential role for artificial intelligence in supporting decision-making based on the results of surveillance analyses (Dixon et al., 2013; Mamiya et al., 2015; Shaban-Nejad et al., 2017). It is not possible to explore the potential contribution to OSH surveillance of all these developments, but two analytical topics of direct relevance to OSH surveillance are highlighted: small-area estimation and aberration detection.

### *Small-Area Estimation*

While several national surveys serve as essential sources of population-based information on work-related injuries, their capacity to "drill down" to specific subnational areas of the nation to facilitate comparable analyses is severely limited by cost constraints. For example, the proposed BLS household survey of occupational injuries and illnesses (see Chapter 5) specified as a supplement to the CPS will help to fill several analytical needs of a comprehensive national surveillance system for occupational safety and health. As currently envisioned by BLS, the proposed sample size

**BOX 7-1**
**Case Study:**
**Coal Workers' Pneumoconiosis Population Surveillance**

Coal workers' pneumoconiosis (CWP) provides an example of where effective surveillance led to public health action. Since 1969, when the Federal Coal Mine Health and Safety law was enacted, there has been continuing reduction in CWP in U.S. coal mines attributed to effective control of dust exposures. Progress has been tracked by NIOSH's Coal Workers' X-ray Surveillance Program (CWXSP) that documented ongoing reduction in CWP late into the 1990s. However, beginning around 2000 the program began to note increased prevalence of early CWP. During the same period a Department of Labor advisory report recommended reducing coal mine dust exposure limits to further eliminate CWP (DOL, 1996). NIOSH decided to analyze the CWXSP in greater depth and found that, among states with coal miners, there was a tenfold difference in proportion of miners that were found to have CWP. Examining all CWP cases, the subset with rapidly progressive CWP were younger than the others and were more likely to have worked in smaller mines ("hot spots"). NIOSH then turned to the Mine Safety and Health Administration (MSHA) records of dust monitoring exposure data—data collected regularly from all underground mines. These, however, did not show evidence for increased exposure. Exploring further to understand why dust levels were not associated with CWP findings, NIOSH made several observations: increased production (coal prices were high at the time), a concomitant change in work schedules (12-hour work days), exposure controls lagging technology advances especially in long-wall mining, a monitoring rule that called for sampling only 8 of the 12 hours (production and, hence, exposure could have been higher later in work shifts due to changes during shift), and among the smaller mines successful dust control was more fragile with shorter lifespans due to market forces. All in all, NIOSH concluded that the hot spots indicated a need for improvements in prevention of CWP. This and other NIOSH surveillance data provided an important impetus for MSHA to promulgate a new Coal Mine Dust

ranging from 51,000 to 57,000 individuals was specified to simultaneously satisfy the multiple objectives of sample representativeness, data quality, timeliness, and cost. While such sample specifications achieve solid levels of precision at the national level, they may result in imprecise estimates at the county or sub-county levels. There are also confidentiality constraints imposed on the release of such subnational geographic content on analytic files made available to the public.

The capacity to obtain reliable small-area estimates derived from national survey data can be substantially enhanced by application of small-area estimation methods. These techniques combine available sample data with auxiliary data using model relationships to improve the reliability

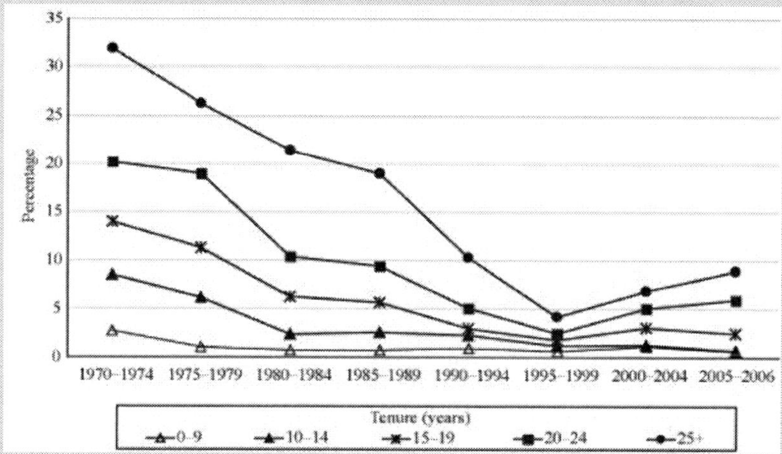


CWXSP: Percentage of Examined Miners with Coal Workers' Pneumonoconiosis (category 1/0) by Tenure in Mining, 1970-2006.

Rule in 2014 to improve protections against respiratory disease caused by coal mine dust. These included improved exposure assessment using new technology (a real-time Continuous Personal Dust Monitor), a reduction in the permissible exposure limit for coal mine dust, and expanded medical surveillance to include surface miners and spirometry. Because CWP is a chronic condition and development of CWP is a lagging indicator, the effect of implementing the new rule may not be evident in CWXSP data tracking for a number of years.

SOURCES: NIOSH, 1995, 2011; Antao et al., 2015.

of the resulting estimates (Pfeffermann, 2013; Vaish et al., 2013; Folsom and Vaish, 2014). Effective small-area estimation methodology depends on the availability of useful predictors reasonably related to the outcome measures. These predictors are annually obtained from various sources and federal agencies such as the U.S. Census Bureau and the American Community Survey.

### *Aberration Detection*

A concept related to small-area analysis is aberration detection, which refers to the detection of statistical anomalies, or aberrations, in surveillance data. Aberrations can be sought along any dimension of the data, but most commonly, deviations from expectation are sought for risk factors or outcomes across time, geographical location, including workplace, and personal attributes, including occupation. As with small-area analysis, it can be challenging to reliably detect true aberrations if small amounts of data are divided into many categories, as many of the resulting cells are likely to contain a small number of events.

Historically, the tendency in public health surveillance has been to search for aberrations in time, aggregating cases to estimate rates, and then applying statistical time-series methods to the rates to detect the onset of an infectious disease epidemic or the effect of an environmental exposure in the whole population, or in subpopulations through stratified analyses. This approach remains a useful strategy as was demonstrated recently for coal dust exposure (see the case study in Box 7-1 on population surveillance). However, advances in statistics, artificial intelligence, and computing power have made it possible to automate the routine analysis of large volumes of individual-level data to detect changes across multiple aspects of time, geography, and personal attributes (Lombardo and Buckeridge, 2007). Had this type of analytical capacity been in place in 1990, the "hot spots" problem described in Box 7-1 may have been identified and results acted on a decade earlier.

Advances in analytical methods have been motivated by access to large volumes of case-level data for surveillance, for example from clinical information systems such as emergency department triage systems and EHRs. A wide range of machine learning methods have been used to detect cases in surveillance data and to identify unusual patterns among cases. More recently, researchers have developed methods to analyze temporal patterns within patient trajectories and then identify unusual subpopulations requiring closer inspection (Lange et al., 2015). There has also been progress in developing methods to integrate data from multiple sources, for example explicitly linking multiple health outcomes to measures of exposure (Morrison et al., 2016).

In occupational health, computationally intensive approaches to aberration detection could support accurate detection of unusual increases in injury or disease among types of occupations in specific workplaces (Kulldorff et al., 2003). The challenges are substantial, especially when large amounts of heterogeneous data are involved, some of which may be sought from commercial entities with proprietary interests or from entities that may offer political or privacy objections to the data use. Despite these challenges, as is done in other types of public health surveillance, OSH surveillance systems could take advantage of advanced statistical and machine learning methods together with data processing methods to automatically analyze OSH data as they are collected.

There is currently no routine review by federal or state agencies of individual, case-level data collected by various OSH surveillance systems to identify new hazards or the introduction of previously recognized hazards in new industries. NIOSH, through an interagency agreement, has access to case-based data in the BLS CFOI database, but it lacks the resources to routinely access these data. As noted earlier in this report, NIOSH does not have access to the SOII data. OSHA has a database of inspections, including fatalities, but again, it does not have the resources to routinely review these data. Without routine analysis of these data, important opportunities to identify significant concerns are missed. For example, only as a special project in 2013, after a request from a state that identified three deaths, did OSHA review its fatality investigation database and identify there had been 10 additional deaths among bathtub refinishers from 2001 to 2011 (Chester et al., 2012; see Box 7-2).

The inclusion of routine rapid analysis of case-level data as a component of the envisioned 21st-century OSH surveillance system would permit more timely identification of emergent OSH injuries, illnesses, and exposures and help facilitate concomitant rapid responses and interventions. Such coordinated analysis within a system of systems could help to identify emergent occupational illnesses and hazards that currently go unnoticed due to the lack of integration across surveillance systems and the inability to analyze data in real time.

**Recommendation N: To identify emerging and serious OSH injuries, illnesses, and exposures in a timely fashion, NIOSH (in coordination with OSHA, BLS, and the states) should develop and implement a plan for routine, coordinated, rapid analysis of case-level OSH data collected by different surveillance systems, followed by the timely sharing of the findings.**

In the near term:

- Develop analytical objectives, identifying the outcomes that would benefit from routine, rapid analysis and continuous monitoring across OSH surveillance systems; and
- Review technical and legal strategies for conducting analyses, including novel analytical methods and strategies for distributed analysis and ongoing analysis as the data evolve over time.

In the longer term:

- Implement routine processes for rapid data analysis, including protocols to guide the interpretation of aberrations in surveillance data.

**BOX 7-2**
**Case Study: Bathtub Refinishers Case-Based Surveillance**

The Michigan Fatality Assessment and Control Evaluation (MIFACE) investigated the death of a 52-year-old male bathtub refinisher. The MIFACE program investigates occupational fatalities to identify preventable risks. In this case, the individual had been stripping a bathtub in an apartment bathroom using a commercial agent that contained >60% methylene chloride (MC) as the active ingredient. The product used was sold to strip paint from aircraft. The work was done with no ventilation and without use of a supplied air respirator that could have prevented the death.

The MIFACE program reviewed all Michigan occupational fatalities in recent years and identified two more deaths among workers performing similar tasks. Staff members notified NIOSH which in turn notified OSHA. A review of OSHA's fatality inspections for the previous 10 years identified 10 more fatalities from nine different states where fatalities were attributed to the same cause. Six different commercial agents had been used, none of which mentioned on the label the hazards of using the product for bathtub refinishing. Two of the three Michigan cases were found in the OSHA records, for a total of 12 death investigations, but the third Michigan death was not mentioned since that worker was self-employed and outside the jurisdiction of OSHA.

The MIFACE program issued a 21-page investigation report and a one-page hazard alert, and published an article in *Morbidity and Mortality Weekly Report* in collaboration with NIOSH and OSHA. Recommendations were made that methylene chloride products should not be used to perform bathtub refinishing.

Surveillance worked where it was applied. But OSHA had no one examining patterns in their fatality inspection records so the recognition of MC as a common element was not discovered until MIFACE raised a concern. Recognition might have occurred even earlier if OSHA Integrated Management Information System exposure data had been reviewed. The state OSHA program in Washington reported methylene chloride levels 17 times higher than the Permissible Exposure Limit in 2005.

If a national alert system had been operating with a plan for ongoing analysis of records from OSHA and other sources, the problem may well have been discovered sooner and fatalities prevented.

SOURCES: Lofgren et al., 2010; Chester et al., 2012.

## Dissemination of Surveillance Information

Although the process of disseminating information to guide public health actions is critical to realizing the potential benefits of surveillance, this process does not always receive the attention it deserves. As with any communication strategy, dissemination of surveillance information is likely to be most effective if the audience is identified and structured into segments, and then each audience segment is targeted using appropriate media and messages. New information technologies can aid greatly in this regard as most people are now instantly accessible via mobile devices and technologies such as social media and the EHR allow messages to be tailored to specific audiences and contexts.

### *Mobile Devices*

The ubiquitous nature of mobile devices makes them well suited to disseminating knowledge and information. Guidelines and evidence regarding risks and preventive measures are easily accessible using mobile devices. Information obtained from analyzing crowdsourced data on exposure risks could also be pushed to employers and employees, indicating nearby risks, as has been done with infectious disease exposures (HealthMap, 2017).

### *Social Media*

As mentioned earlier, social media data have been used as a source of surveillance, for example, to detect the onset of a seasonal influenza epidemic by analyzing the frequency with which influenza symptoms are mentioned. However, social media also provide an opportunity for public health authorities to engage with people by disseminating targeted information, which may help to prevent illness or allow identification of ongoing threats to health.

In the context of influenza surveillance, researchers have used social media to notify individuals at risk of disease where influenza vaccine is available nearby (Smolinski et al., 2015). This type of feedback combines information about patient risk, patient location, and the location of prevention resources available nearby (i.e., vaccine clinics) to increase the use of evidence-based preventive maneuvers, with the aim of preventing disease. Similarly, in the context of foodborne disease, public health agencies have used automated software to identify people making posts about being ill after visiting a restaurant. These people then receive an automated message asking them to access a website and provide further details about their experience (Harris et al., 2014).

These examples have direct analogies to occupational health. For example, people at higher risk of an occupational injury or disease could be directed toward resources, that may allow them to prevent an injury or disease such as emphasizing the use of available protection when working at heights. Automated bots could be used to flag people who may be commenting on an occupational injury of disease. Once identified, it would be possible to direct those people to a website where they could provide further data, and analysis of those data could trigger an assessment of their workplace.

### *Electronic Health Records*

While electronic health records are a technology that can greatly facilitate the capture of data, they also present an opportunity for disseminating information produced through the analysis of surveillance data. The dissemination of surveillance information via an EHR can take different forms, including through alerts to identify patients who may be presenting with an occupational injury or disease and by providing feedback to health care providers regarding their management of patients with occupational injury or disease.

The ability to provide alerts to clinicians through an EHR has been demonstrated for communicable disease control. In that context, knowledge of an increase in infectious disease activity in a geographic region has been used to alert physicians when a patient presents from the same geographic region with symptoms consistent with the infections disease in question (Lurio et al., 2010). Similarly, researchers have shown that in emergency department encounters for some infectious disease, accounting for the prevalence of the infectious disease ascertained from surveillance data can enhance clinical decision rules, allowing more accurate diagnosis (Fine et al., 2007). Both examples have analogous applications for OSH surveillance. For example, if public health authorities recognize a cluster of illness or injury associated with an occupation or workplace, then alerts could be constructed to prompt physicians to consider an occupation etiology when patients present with a similar illness or injury. The feasibility of disseminating information in this manner was demonstrated in Massachusetts when NIOSH funded the state health department to incorporate data on occupation into electronic heath records for a major health care system in the state. The health department coded data for 26,000 patients and entered these data into the system. Based on occupation patterns alone, it was noted that a high proportion of Portuguese-speaking women were house cleaners, and Spanish-speaking men were painters. Consequently, multilingual materials on occupation-specific hazards and controls were made available on the system so that clinicians could distribute them to patients in these jobs (Brightman et al., 2013).

In some clinical settings, data captured through EHRs are pooled, analyzed to quantify variations in care across providers, and the results are then fed back to clinicians to help them situate their practice pattern in relation to their peers and clinical practice guidelines. For example, in some primary care settings, data from EHRs are collected and analyzed to determine the proportion of patients with type 2 diabetes who have had a hemoglobin A1C test performed recently. This information is then fed back to each participating primary care provider, allowing them to identify their management of such patients in relation to their peers (Seitz et al., 2011). In occupational health, a similar strategy could be employed, for example, to provide feedback to physicians regarding their management of occupational injuries or disease.

*Rapid Alert Networks*

As in other areas of public health surveillance, an early warning alert network could use any or all of the strategies described above to disseminate important findings to the OSH community at large. Doing so could promote fast-track situational awareness of emergent occupational illnesses and hazards, accelerate more focused analyses to determine the level of imminent risk, and stimulate decisions on prompt responses and interventions to mitigate the danger. Responses might include further targeted surveillance, OSH community messaging, and more focused research investigations making greater use of other relevant available data sources.

**Recommendation O: To promote and facilitate the use of surveillance information for prevention, and to present more comprehensive information on the extent, distribution, and characteristics of OSH injuries, illnesses, and exposures, NIOSH (in coordination with and input from OSHA, BLS, and the states) should establish a coordinated strategy and mechanism for timely dissemination of surveillance information.**

In the near term:

- Clarify target populations for different types of surveillance information (e.g., rapid alerts, trends, etc.);
- Establish a plan for accessing, integrating, and disseminating information from different surveillance sources; and
- Develop policies and criteria to address individuals' and employers' privacy and confidentiality considerations through a process that provides for stakeholder input and includes privacy experts in the development of these policies and in the design of surveillance systems.

In the longer term:

- Implement a coordinated information dissemination strategy, making use of different technologies as appropriate to communicate information to those who need it to take action for prevention.

This coordination would augment and not replace the activities and authority of individual agencies. An overall dissemination strategy will provide a better understanding of occupational injuries and illnesses to assist in the prioritization and evaluation of prevention activity.

## ENHANCE TRAINING AND SUPPORT FOR OSH SURVEILLANCE PRACTITIONERS

Although a surveillance system is often thought of as a technical system, it is better conceptualized as a sociotechnical system, which is as dependent on skilled people as it is on technical components. A range of individuals is necessary to establish and effectively operate surveillance systems, but the key disciplines that contribute to the science (Thacker et al., 1989) and practice of OSH surveillance through the effective application of new information technologies are epidemiology, biostatistics, and biomedical informatics.

Trainees in OSH are likely to receive instruction in epidemiology and biostatistics, although rarely are they taught to apply methods from these disciplines to surveillance. For example, designs for evaluating surveillance systems and statistical methods for aberration detection are not taught routinely in OSH or other public health programs. The situation is more concerning for biomedical informatics. This discipline is identified as a core public health competency (CLBAPHP, 2014), however, it is taught to varying degrees across education institutions, and the specific aspects of informatics relevant to OSH surveillance are not taught routinely in many academic institutions.

**Recommendation P: NIOSH, OSHA, and BLS should work together to encourage education and training of the surveillance workforce in disciplines necessary for developing and using surveillance systems, including epidemiology, biomedical informatics, and biostatistics.**

In the near term:

- Identify the core competencies required for OSH surveillance and promote the science of surveillance;
- Review the curricula of existing surveillance courses;

- Collaborate with educational organizations to establish or modify training programs accordingly; and
- Require surveillance courses in all funded training programs, especially in the Education and Research Center and Program Project training grants.

In the longer term:

- Contribute to development of surveillance courses and conferences that provide training in surveillance methods.

## REFERENCES

Antao, V. C., E. L. Personk, L. Z. Sokolow, A. L. Wolfe, G. A. Pinheiro, J. M. Hale, and M. D. Attfield. 2005. Rapidly progressive coal workers' pneumoconiosis. *Occupational & Environmental Medicine* 62:670-674.

Banke-Thomas, A. O., B. Madaj, A. Charles, and N. van den Broek. 2015. Social Return on Investment (SROI) methodology to account for value for money of public health interventions: A systematic review. *BMC Public Health* 15(1):582.

BLS (Bureau of Labor Statistics). 2015. The quest for meaningful and accurate occupational health and safety statistics. *Monthly Labor Review*. Available online at https://www.bls.gov/opub/mlr/2015/article/the-quest-for-meaningful-and-accurate-occupational-health-and-safety-statistics.htm (accessed July 18, 2017).

BLS. 2017a. *Data Users Advisory Committee*. Available online at https://www.bls.gov/advisory/duac.htm (accessed June 23, 2017).

BLS. 2017b. *Census of Fatal Occupational Injuries (CFOI)—Current and revised data*. Available online at https://www.bls.gov/iif/oshcfoi1.htm (accessed April 4, 2017).

Brabham, D. C., K. M. Ribisl, T. R. Kirchner, and J. M. Bernhardt. 2014. Crowdsourcing applications for public health. *American Journal of Preventive Medicine* 46(2):179-187.

Brauch, R. 2015. How technology megatrends are shaping the future of safety, health, and environmental monitoring. *Occupational Health and Safety* 84(5):34-36, 38.

Brightman, L., K. Souza, L. S. Azaroff, L. Davis, and R. Goldman. 2013. *Collection and recording of patient occupation information in the EHR: A pilot project in a primary care setting*. Presented at the 141st APHA Annual Meeting, November 4, 2013. Boston, MA. Available online at https://apha.confex.com/apha/141am/webprogramadapt/Paper288596.html (accessed July 18, 2017).

Burstyn, I., A. Slutsky, D. G. Lee, A. B. Singer, Y. An, and Y. L. Michael. 2014. Beyond crosswalks: Reliability of exposure assessment following automated coding of free-text job descriptions for occupational epidemiology. *Annals of Occupational Hygiene* 58(4):482-492.

CDC (Centers for Disease Control and Prevention). 2001. Updated guidelines for evaluating public health surveillance systems: Recommendations from the guidelines working group. *Morbidity and Mortality Weekly Report* 50(RR-13):13-24.

Chester, D., K. D. Rosenman, G. R. Grimes, K. Fagan, and D. N. Castillo. 2012. Fatal exposure to methylene chloride among bathtub refinishers-United States, 2000-2011. *Morbidity and Mortality Weekly Report* 61(7):119-122.

Chew, C., and G. Eysenbach. 2010. Pandemics in the age of Twitter: Content analysis of tweets during the 2009 H1N1 outbreak. *PLoS ONE* 5(11):e14118.

CLBAPHP (The Council on Linkages Between Academia and Public Health Practice). 2014. *Core competencies for public health professionals*. Available online at http://www.phf.org/resourcestools/Documents/Core_Competencies_for_Public_Health_Professionals_2014June.pdf (accessed July 18, 2017).

Cooper, B. S., and D. P. Rice. 1976. The economic cost of illness revisited. *Social Security Bulletin* 39(2):21-36.

Des Jarlais, D. C., C. Lyles, and N. Crepaz. 2004. Improving the reporting quality of nonrandomized evaluations of behavioral and public health interventions: The TREND statement. *American Journal of Public Health* 94(3):361-366.

Dixon, B. E., R. E. Gamache, and S. J. Grannis. 2013. Towards public health decision support: A systematic review of bidirectional communication approaches. *Journal of the American Medical Informatics Association* 20(3):577-583.

DOL (U.S. Department of Labor). 1996. *Report of the Secretary of Labor's Advisory Committee on the Elimination of Pneumoconiosis among Coal Miner Workers*, October. Washington, DC: Mine Safety and Health Administration.

Donaire-Gonzalez, D., A. Valentín, A. de Nazelle, A. Ambros, G. Carrasco-Turigas, E. Seto, M. Jerrett, and M. J. Nieuwenhuijsen. 2016. Benefits of mobile phone technology for personal environmental monitoring. *JMIR mHealth and uHealth* 4(4):e126.

Edmunds, M., L. Thorpe, M. Sepulveda, C. Bezold, and D. A. Ross. 2014. The future of public health informatics: Alternative scenarios and recommended strategies. *eGEMS (Washington, DC)* 2(4):1156. Available online at https://www.ncbi.nlm.nih.gov/pmc/articles/PMC4371422 (accessed July 18, 2017).

EPA (U.S. Environmental Protection Agency). 2010. *Valuing Mortality Risk Reductions for Environmental Policy: A White Paper*. Washington, DC: EPA.

EU-OSHA (European Agency for Safety and Health at Work). 2017. *Estimating the Cost of Work-Related Accidents and Ill-Health: An Analysis of European Data Sources*. Luxembourg: Publications Office of the European Union.

Evans, D. E., B. K. Ku, M. E. Birch, and K. H. Dunn. 2010. Aerosol monitoring during carbon nanofiber production: Mobile direct-reading sampling. *Annals of Occupational Hygiene* 54(5):514-531.

Fine, A. M., L. E. Nigrovic, B. Y. Reis, E. F. Cook, and K. D. Mandl. 2007. Linking surveillance to action: Incorporation of real-time regional data into a medical decision rule. *Journal of the American Medical Informatics Association* 14(2):206-211.

Fisher, D. A. 2006. *An Emergent Perspective on Interoperation in Systems of Systems*. Pittsburgh, PA: Carnegie Mellon Software Engineering Institute.

Folsom, R. E., and A. K. Vaish. 2014. Robust small area estimation with additive random coefficients (ARC) models. Frontiers of Hierarchical Modeling in Observational Studies, Complex Surveys and Big Data Conference Honoring Professor Malay Ghosh, College Park, MD.

Freifeld, C. C., R. Chunara, S. R. Mekaru, E. H. Chan, T. Kass-Hout, A. Ayala Iacucci, and J. S. Brownstein. 2010. Participatory epidemiology: Use of mobile phones for community-based health reporting. *PLoS Medicine* 7(12):e1000376.

Fricker, R. 2013. *Introduction to Statistical Methods for Biosurveillance*. Cambridge, UK: Cambridge University Press.

Gillum, L. A., C. Gouveia, E. R. Dorsey, M. Pletcher, C. D. Mathers, C. E. McCulloch, and S. C. Johnston. 2011. NIH disease funding levels and burden of disease. *PLoS ONE* 6(2):e16837.

Gini, R., M. Schuemie, J. Brown, P. Ryan, E. Vacchi, M. Coppola, W. Cazzola, P. Coloma, R. Berni, G. Diallo, J. L. Oliveira, P. Avillach, G. Trifirò, P. Rijnbeek, M. Bellentani, J. van Der Lei, N. Klazinga, and M. Sturkenboom. 2016. Data extraction and management in networks of observational health care databases for scientific research: A comparison of EU-ADR, OMOP, Mini-Sentinel and MATRICE strategies. *eGEMS (Washington, DC)* 4(1):1189.

Gold, M. R., J. E. Siegel, L. B. Russell, and M. C. Weinstein. 1996. *Cost-Effectiveness in Health and Medicine*. New York: Oxford University Press.

Gross, C. P., G. F. Anderson, and N. R. Powe. 1996. The relation between funding by the National Institutes of Health and the burden of disease. *New England Journal of Medicine* 340(24):1881-1887.

Haddix, A. C., S. M. Teutsch, P. A. Shaffer, and D. O. Dunet. 1996. *Prevention Effectiveness: A Guide to Decision Analysis and Economic Evaluation*. New York: Oxford University Press.

Hanser, S., A. Zaiss, and S. Schulz. 2009. Health care procedures. *Methods of Information in Medicine* 48(6):540-545.

Harber, P., and G. Leroy. 2017. Feasibility and utility of lexical analysis for occupational health data. *Journal of Occupational and Environmental Medicine* 59(6):578-587.

Harris, J. K., R. Mansour, B. Choucair, J. Olson, C. Nissen, and J. Bhatt. 2014. Health department use of social media to identify foodborne illness - Chicago, Illinois, 2013-2014. *Morbidity and Mortality Weekly Report* 63(32):681-685.

HealthMap. 2017. *Outbreaks near me*. Available online at http://www.healthmap.org/outbreaksnearme (accessed July 18, 2017).

HSE (Health and Safety Executive). 2016. *Health and safety at work*. Available online at http://www.hse.gov.uk/statistics/overall/hssh1516.pdf?pdf=hssh1516 (accessed May 8, 2017).

Keating, C., R. Rogers, R. Unal, D. Dryer, A. Souza-Poza, R. Safford, W. Peterson, and G. Rabadi. 2003. System of systems engineering. *Engineering Management Journal* 15(3):36-45.

Koeman, T., N. S. Offermans, Y. Christopher-de Vries, P. Slottje, P. A. Van Den Brandt, R. A. Goldbohm, H. Kromhout, and R. Q. Vermeulen. 2013. JEMs and incompatible occupational coding systems: Effect of manual and automatic recoding of job codes on exposure assignment. *Annals of Occupational Hygiene* 57(1):107-114.

Kulikowski C. A., E. H. Shortliffe, L. M. Currie, P. L. Elkin, L. E. Hunter, T. R. Johnson, I. J. Kalet, L. A. Lenert, M. A. Musen, J. G. Ozbolt, J. W. Smith, P. Z. Tarczy-Hornoch, and J. J. Williamson. 2012. AMIA Board white paper: Definition of biomedical informatics and specification of core competencies for graduate education in the discipline. *Journal of the American Medical Informatics Association* 19:931–938.

Kulldorff, M., Z. Fang, and S. J. Walsh. 2003. A tree-based scan statistic for database disease surveillance. *Biometrics* 59(2):323-331.

Lange, J. M., R. A. Hubbard, L. Y. Inoue, and V. N. Minin. 2015. A joint model for multistate disease processes and random informative observation times, with applications to electronic medical records data. *Biometrics* 71(1):90-101.

Leigh, J. P. 2011. Economic burden of occupational injury and illness in the United States. *Milbank Quarterly* 89(4):728-772.

Lofgren, D. J., C. K. Reeb-Whitaker, and D. Adams. 2010. Surveillance of Washington OSHA exposure data to identify uncharacterized or emerging occupational health hazards. *Journal of Occupational and Environmental Hygiene* 7:375-388.

Lombardo, J. S., and D. L. Buckeridge, eds. 2007. *Disease Surveillance: A Public Health Informatics Approach*. Hoboken, NJ: John Wiley & Sons.

Luo, F., and C. Florence. 2017. State-level lifetime medical and work-loss costs of fatal injuries—United States, 2014. *Morbidity and Mortality Weekly Report* 66(1):1-11.

Lurio, J., F. P. Morrison, M. Pichardo, R. Berg, M. D. Buck, W. Wu, K. Kitson, F. Mostashari, and N. Calman. 2010. Using electronic health record alerts to provide public health situational awareness to clinicians. *Journal of the American Medical Informatics Association* 17(2):217-219.

Mac Kenzie, W. R., A. J. Davidson, A. Wiesenthal, J. P. Engel, K. Turner, L. Conn, S. J. Becker, S. Moffatt, S. L. Groseclose, J. Jellison, J. Stinn, N. Y. Garrett, L. Helmus, B. Harmon, C. L. Richards, J. R. Lumpkin, and M. F. Iademarco. 2016. The promise of electronic case reporting. *Public Health Reports* 131(6):742-746.

Maier, M. W. 1998. Architecting principles for systems-of-systems. *Systems Engineering* 1(4):267-284.

Mamiya, H., K. Schwartzman, A. Verma, C. Jauvin, M. Behr, and D. Buckeridge. 2015. Towards probabilistic decision support in public health practice: Predicting recent transmission of tuberculosis from patient attributes. *Journal of Biomedical Informatics* 53: 237-242.

Mandl, K. D., J. M. Overhage, M. M. Wagner, W. B. Lober, P. Sebastiani, F. Mostashari, J. A. Pavlin, P. H. Gesteland, T. Treadwell, E. Koski, L. Hutwagner, D. L. Buckeridge, R. D. Aller, and S. Grannis. 2004. Implementing syndromic surveillance: A practical guide informed by the early experience. *Journal of the American Medical Informatics Association* 11(2):141-150.

Michie, S., M. Richardson, M. Johnston, C. Abraham, J. Francis, W. Hardeman, M. P. Eccles, J. Cane, and C. E. Wood. 2013. The behavior change technique taxonomy (v1) of 93 hierarchically clustered techniques: Building an international consensus for the reporting of behavior change interventions. *Annals of Bevioural Medicine* 46(1):81-95.

Mitchell, L., and J. V. Ross. 2016. A data-driven model for influenza transmission incorporating media effects. *Royal Society Open Science* 3(10):160481.

Morrison, K. T., G. Shaddick, S. B. Henderson, and D. L. Buckeridge. 2016. A latent process model for forecasting multiple time series in environmental public health surveillance. *Statistics in Medicine* 35(18):3085-3100.

Muennig, P. 2008. *Cost-Effectiveness Analyses in Health: A Practical Approach*, 2nd Ed. San Francisco, CA: Jossey-Bass.

NIOSH (National Institute for Occupational Safety and Health). 1995. *Criteria for a Recommended Standard: Occupational Exposure to Coal Mine Dust.* DHHS (NIOSH) Publication No. 95-106. Available online at https://www.cdc.gov/niosh/docs/95-106/pdfs/95-106.pdf (accessed December 6, 2017).

NIOSH. 2011. Coal mine dust exposures and associated health outcomes: A review of information published since 1995. *Current Intelligence Bulletin 64.* DHHS (NIOSH) Publication No. 2011-172. Available online at https://www.cdc.gov/niosh/docs/2011-172/pdfs/2011-172.pdf (accessed December 6, 2017).

NIOSH. 2016. *Center for Direct Reading and Sensor Technologies.* Available online at https://www.cdc.gov/niosh/docs/2016-132/ (accessed July 18, 2017).

NIOSH.2017a. *Worker health surveillance.* Available online at https://www.cdc.gov/niosh/topics/surveillance/default.html (accessed March 20, 2017).

NIOSH. 2017b. *Coal workers' health surveillance program.* Available online at https://www.cdc.gov/niosh/topics/cwhsp/ (accessed June 4, 2017).

NIOSH. 2017c. *Fatality Assessment and Control Evaluation (FACE) program.* Available online at https://www.cdc.gov/niosh/face/default.html (accessed April 4, 2017).

NIOSH. 2017d. *Firefighter fatality investigation and prevention.* Available online at https://www.cdc.gov/niosh/fire/default.html (accessed January 29, 2017).

NIOSH. 2017e. *NIOSH and Partners Work to Prevent Worker Deaths from Exposures to Hydrocarbon Gases and Vapors at Oil and Gas Wellsites.* DHHS (NIOSH) Publication No. 2017-110. Available online at https://www.cdc.gov/niosh/docs/2017-110 (accessed May 8, 2017).

NIOSH. 2017f. *Welcome to eWORLD.* Available online at https://wwwn.cdc.gov/eworld (accessed June 14, 2017).

NIOSH. 2017g. *Bloodborne infectious diseases: HIV/AIDS, hepatitis B, hepatitis C.* Available online at https://www.cdc.gov/niosh/topics/bbp/ (accessed May 8, 2017).

NIOSH. 2017h. *Health and safety practices survey of healthcare workers.* Available online at https://www.cdc.gov/niosh/topics/healthcarehsps/aboutstudy.html (accessed June 14, 2017).

Olson, D. R., M. Paladini, W. B. Lober, and D. L. Buckeridge and ISDS Distribute Working Group. 2011. Applying a new model for sharing population health data to national syndromic influenza surveillance: DiSTRIBuTE project proof of concept, 2006 to 2009. *PLoS Currents.* 3:RRN 1251. Available online at http://currents.plos.org/influenza/article/applying-a-new-model-for-sharing-261w1jjdm6zrb-5 (accessed July 18, 2017).

OSHA (Occupational Safety and Health Administration). 2017. *Data and statistics.* Available online at https://www.osha.gov/oshstats/index.html (accessed July 18, 2017).

Patel, M. D., K. M. Rose, C. R. Owens, H. Bang, and J. S. Kaufman. 2012. Performance of automated and manual coding systems for occupational data: A case study of historical records. *American Journal of Industrial Medicine* 55(3):228-231.

Pew Research Center. 2015. *U.S. smartphone use in 2015.* Available online at http://www.pewinternet.org/2015/04/01/us-smartphone-use-in-2015 (accessed July 18, 2017).

Pfeffermann, D. 2013. New important developments in small area estimation. *Statistical Science* 28(1):40-68.

Pirnstill, C. W., and G. L. Coté. 2015. Malaria diagnosis using a mobile phone polarized microscope. *Scientific Reports* 5:13368.

Pocock, M. J., J. C. Tweddle, J. Savage, L. D. Robinson, and H. E. Roy. 2017. The diversity and evolution of ecological and environmental citizen science. *PLoS ONE* 12(4):e0172579.

Popper, S., S. Bankes, R. Callaway, and D. DeLaurentis. 2004. *System-of-Systems Symposium: Report on a Summer Conversation, July 21-22, 2004.* Potomac Institute for Policy Studies, Arlington, VA.

Powell, G. E., H. A. Seifert, T. Reblin, P. J. Burstein, J. Blowers, J. A. Menius, J. L. Painter, M. Thomas, C. E. Pierce, H. W. Rodriguez, J. S. Brownstein, C. C. Freifeld, H. G. Bell, and N. Dasgupta. 2016. Social media listening for routine post-marketing safety surveillance. *Drug Safety* 39(5):443-454.

Rice, D. P. 1967. Estimating the cost of illness. *American Journal of Public Health* 57(3):424-440.

Rice, D. P., T. A. Hodgson, and A. N. Kopstein. 1985. The economic costs of illness: A replication and update. *Health Care Financing Review* 7(1):61-80.

Schulte, P. A., R. Pana-Cryan, T. Schnorr, A. L. Schill, R. Guerin, S. Felknor, and G. R. Wagner. 2017. An approach to assess the burden of work-related injury, disease, and distress. *American Journal of Public Health* 107(7):1051-1057.

Seitz, P., T. Rosemann, J. Gensichen, and C. A. Huber. 2011. Interventions in primary care to improve cardiovascular risk factors and glycated haemoglobin (HbA1c) levels in patients with diabetes: A systematic review. *Diabetes Obesity and Metabolism* 13(6):479-489.

Shaban-Nejad, A., M. Lavigne, A. Okhmatovskaia, and D. L. Buckeridge. 2017. PopHR: A knowledge-based platform to support integration, analysis, and visualization of population health data. *Annals of the New York Academy of Sciences* 1387(1):44-53.

Shortliffe, E. H. 2014. Biomedical informatics: The science and the pragmatics. Pp. 3-37 in *Biomedical Informatics: Computer Applications in Health Care and Biomedicine*, E. H. Shortliffe and J. J. Cimino, eds. London: Springer.

Smolinski, M. S., A. W. Crawley, K. Baltrusaitis, R. Chunara, J. M. Olsen, O. Wójcik, M. Santillana, A. Nguyen, and J. S. Brownstein. 2015. Flu near you: Crowdsourced symptom reporting spanning 2 influenza seasons. *American Journal of Public Health* 105(10):2124-2130.

Steenland, K., C. Burnett, N. Lalich, E. Ward, and J. Hurrel. 2003. Dying for work: The magnitude of US mortality from selected causes of death associated with occupation. *American Journal of Industrial Medicine* 43:461-482.

Thacker, S. B., R. L. Berkelman, and D. F. Stroup. 1989. The science of public health surveillance. *Journal of Public Health Policy* 10(2):187-203.

Thorpe, K. E., L. Allen, and P. Joski. 2015. The role of chronic disease, obesity, and improved treatment and detection in accounting for the rise in healthcare spending between 1987 and 2011. *Applied Health Economics and Health Policy* 13(4):381-387.

Vaish, A. K., R. E. Folsom, K. Spagnola, N. Sathe, and A. K. Hughes. 2013. An empirical study to evaluate the performance of synthetic estimates of substance use in the National Survey on Drug Use and Health. Pp. 2538-2547 in *Proceedings of the American Statistical Association, Section on Survey Research Methods*. Alexandria, VA: American Statistical Association.

Vogel, J., J. S. Brown, T. Land, R. Platt, and M. Klompas. 2014. MDPHnet: Secure, distributed sharing of electronic health record data for public health surveillance, evaluation, and planning. *American Journal of Public Health* 104(12):2265-2270.

Wells, G. D., and A. P. Sage. 2008. Engineering of a system of systems. Pp. 44-76 in *Systems of Systems Engineering: Principles and Applications*, M. Jamshidi, ed. Boca Raton, FL: CRC Press.

Welvaert, M., and P. Caley. 2016. Citizen surveillance for environmental monitoring: Combining the efforts of citizen science and crowdsourcing in a quantitative data framework. *SpringerPlus* 5(1):1890.

# 8

# Next Steps for Improving Worker Safety and Health Through a Smarter Occupational Surveillance System

The Occupational Safety and Health Act of 1970 sought "to assure so far as possible every working man and woman in the Nation safe and healthful working conditions and to preserve our human resources." Progress toward this goal has been steady, but the human and economic toll from work injury and disease remains unacceptably high. Occupational health surveillance provides essential information necessary to understand the distribution and determinants of the burden of injury and disease among workers, to track changes in this burden over time, and to assess the impact and success of interventions designed to reduce or eliminate the adverse consequences of work across the full spectrum of employment in the United States.

Although there have been advances in occupational safety and health (OSH) surveillance since the 1987 National Research Council (NRC) report, greater advances can be made through the application of new technologies, systems approaches, and coordinated efforts. This final chapter collects the recommended components of an action plan to move toward a smarter and more dynamic OSH surveillance system. It begins by providing one final meta-recommendation that the committee considers essential to achieve a smarter system. Then, drawing on the recommendations made throughout the report that delineate near- and longer-term steps toward progress in OSH surveillance, the chapter outlines the report's call for four major categories of action. It accordingly summarizes the committee's recommendations thematically, rather than following their order of presentation in earlier chapters, with the four major categories as follows:

- Prioritize and coordinate OSH surveillance
- Improve data collection
- Expand biomedical informatics use and capabilities
- Strengthen data analysis and information dissemination for prevention

The discussion refers to the recommendations by their letter designation, and the report provides a listing of the full set of recommendations in Appendix C, which cites them in the order of their presentation in the report. Readers may refer as needed to the individual recommendations by letter in Appendix C when they are cited in the remainder of this chapter.

## VISION FOR A SMARTER SYSTEM

The committee's vision for the future of OSH surveillance is a collaborative system of systems. Recognizing the varying mandates and roles of the many relevant stakeholders, the committee believes that it is possible to strengthen the ongoing coordination and data sharing across federal agencies, between federal and state agencies, across state agencies (e.g., labor and health), and with employers and workers to result in the maximum possible engagement of all. A system of systems approach to OSH surveillance would minimize the undercounting of occupational injuries and illnesses by gathering sufficient data that include nontraditional occupations and worker groups in a representative manner and enhancing prevention-relevant information in surveillance data to include race and ethnicity as well as occupation and industry (general and detailed). It would expand outcomes to include chronic diseases and their causes and include leading indicators, primarily through adequately detailed exposure information. Further, the system of systems would maximize appropriate use of technologies to facilitate all surveillance processes and create structures to disseminate information to levels where it can be acted upon.

There are several critical enabling components that will be leveraged by the agencies responsible for implementing the smarter system. Such efforts begin by extending the capacities of the agencies through targeted enhancements of both existing systems and technical personnel, and through effective communication across agencies. This will be complemented by information and other inputs available through employers, employee representatives, relevant intermediaries, and individuals along with engaged health care systems, all taking full advantage of state-of-the-art technology.

Engagement of employers and health care providers will result in substantive improvements by augmenting existing resources (e.g., effective collaboration with workers' compensation systems, efficient implementation of electronic reporting, including occupational information in electronic health records) along with development of new resources such as voluntary within-industry partnerships to engage collectively in exposure surveillance (e.g., modeled on the Industrial Minerals Association silica exposure assessment in the European Union). At the same time a complementary effort at incorporating leading indicators of risk such as hazard and exposure surveillance will advance with better utilization of existing resources (e.g., Occupational Safety and Health Administration [OSHA] compliance data and National Institute for Occupational Safety and Health [NIOSH] Health Hazard Evaluation information) coupled with enhancements (e.g., collection of required exposure monitoring and industry-specific exposure surveys).

Individuals, in their capacity as workers and beyond, will play an essential role in the smarter system through their participation in population health surveys that incorporate occupational information (e.g., household survey of nonfatal occupational injuries and illnesses [HSOII], the National Health Interview Survey, the Behavioral Risk Factor Surveillance System, and the Medical Expenditure Panel Survey).

Evolving health care systems, along with technology imbedded in the delivery system, will greatly facilitate enriched inputs on health and safety outcome data. Inclusion of occupational information in the electronic health record and advances in health care reporting structures will improve content and efficiency in collecting reports of work-related health conditions.

Fundamental to a successful smarter system will be sufficient and creative use of information technology capacity and resources. These will include effective autocoding of occupational information in all appropriate records, electronic reporting wherever possible from all traditional and emerging reporting sources, and development of hardware and software for efficient collection of information in real time (e.g., exposure and compliance data). To enable the system fully, methods and tools need to be developed for timely and effective collection and analysis of surveillance data. In addition, software needs to be designed and disseminated so that all relevant stakeholders can undertake their own examination of surveillance information so that they can act on findings as quickly as possible to improve health and safety for workers regardless of setting or context.

The committee has based its analysis and recommendations on an understanding that desired improvements in worker safety and health are more likely to be met when an adequate amount of information is available at a level of detail sufficient to facilitate effective public health actions. Producing this information requires coordination of agencies, with input from stakeholders, collection of data that can provide useful information on all components of the workforce, and the ability to process and share information using modern technologies. Even as the committee recognizes that the pathway to achieve a smarter system is neither direct nor without challenging barriers, ultimately, all the elements are attainable in an effectively coordinated national system.

## ACTION STEPS – REPORT RECOMMENDATIONS

This report provides recommendations for improved OSH surveillance, many of which offer both near- and long-term recommended actions. The near-term actions are intended to be possible even when recognizing the constraints on all actors that arise from currently limited resources and complicated historical precedents. The longer-term actions are expected to require new resources (financial and personnel) along with the evolution of elements, some of which will be made possible by implementing shorter-term recommendations while others are beyond the direct control of the leadership for OSH surveillance.

The details of the recommendations are provided throughout the report; the following synthesizes the recommended action steps in the four categories mentioned earlier.

### Prioritize and Coordinate OSH Surveillance

First and foremost, surveillance for occupational health and safety needs to become a priority if it is to serve the core function of providing the information essential to guide public health actions to improve worker safety and health. The committee recognizes that surveillance often exists in the background of public health programs, rising to a level of importance only at times that call for emergency action. It is less well appreciated that, even in the background, the system needs to operate efficiently—seamlessly collecting, collating, and assessing information without interruption to support evidence-based actions, emergency or otherwise. With surveillance as a priority, the development of a centralized coordination of a system of systems can provide the essential evidence to guide prevention efforts that advance program objectives in the most cost-effective manner.

**Recommendation Q (meta-recommendation): The Secretary of Health and Human Services, with the support of the Secretary of Labor, should direct NIOSH to form and lead a coordinating entity in partnership with OSHA, BLS, and other relevant agencies.** The coordinating entity should:

- develop and regularly update a national occupational safety and health surveillance strategic plan that is based on well-articulated objectives;
- coordinate the design and evaluation of an evolving national system of systems for OSH surveillance and for the dissemination of surveillance information provided by these systems;

- publish a report on progress toward the strategic plan's implementation at least every 5 years, documenting advances toward achieving a 21st-century smarter occupational safety and health (OSH) surveillance system; and
- engage partners, including other federal health statistics agencies, state agencies with OSH responsibilities, and stakeholders.

This recommendation is arguably the cornerstone for the advice that the committee is offering. The envisioned coordinating entity is essential if there is to be a cross-agency vision and plan for moving forward, if the other recommendations are to be properly prioritized and carried out, and if the resulting system of systems is to be effectively guided by the principles described in Chapter 2. The evolving system needs to be robust and collaborative, with strong leadership, and needs to use consistent standards across all relevant domains. It needs to assure timely analysis and interpretation of surveillance inputs tied closely with dissemination to relevant actors. And it will need to safeguard privacy and confidentiality, monitor data quality to ensure program efficiency and impact, and be staffed by well-trained public health professionals with access to the tools and technology necessary to achieve surveillance objectives.

The envisioned strategic plan would need to provide an overall vision and framework for OSH surveillance for the nation that government agencies and stakeholders will then implement. As a living document, the OSH surveillance strategic plan can be flexible to adjust to changing priorities, needs, and circumstances. The plan is intended to provide a sustainable framework for accomplishing the core objectives of OSH surveillance:

- Guide immediate action to control threats to occupational health and safety;
- Measure the health and economic burden of work-related injuries or illnesses and monitor trends over time and space;
- Identify industries, occupations, and worksites as well as populations defined by sociodemographic characteristics and work arrangements at high risk for work-related injury, illness, or hazardous exposures;
- Detect and respond to new or emerging workplace hazards or facilitate the investigation of new diseases linked to occupational exposures;
- Guide the planning, implementation, and evaluation of programs and policies intended to prevent and control work-related injuries, illnesses, and hazardous workplace exposures; and
- Generate hypotheses and make pertinent data available for research.

As a system of systems, this smarter system would need to begin by making clear the specific objectives for each of the surveillance systems within the overall system. It would include concrete objectives for each of the key federal agencies. It would also include a detailed plan for engagement of the state-based OSH programs that identifies priority conditions for expanded surveillance, provides guidance on how to use the data generated by the states, and, whenever possible, identifies the agencies (i.e., federal or state) that need to take the lead for these conditions. Resource needs need to be organized, projected, and articulated. Benchmarks need to be identified and realistic timelines need to be specified to keep attention on measurable progress. Furthermore, evaluation needs to be an essential element at all levels, organized so that successful systems can be expanded when possible and replicated at different levels of the organization as appropriate. Systems that do not meet objectives or that cease to be cost effective can be terminated.

In setting forth OHS surveillance as a national priority, the responsible agencies, most centrally NIOSH, will need to organize to make certain that there is a clear line of responsibility and authority over each agency's OSH surveillance activities and personnel. Unless leadership structurally and distinctly recognizes and articulates these actions, the system will likely be incapable of achieving the identified goals. At the same time that those structural changes are being developed, the agencies need to ensure that links across agencies are as seamless as possible and that barriers to timely, efficient sharing of data and information are eliminated. As with most public health activities that address more than one priority, the overall system will need to be founded on a close working relationship between federal and state partners. Together, coordinated federal and state systems offer immense advantages over either operating independently. The report accordingly stresses the value of an effective federal-state partnership and strengthened state efforts, both to facilitate and serve a coordinated national effort to identify and monitor priority conditions and emerging problems, and to foster prevention programs at the level that can best address these concerns (Recommendation C).

The committee also notes that the most effective intervention activities will need to act on the causes and not the consequences of occupational health problems. Exposure and hazard surveillance points the way to primary prevention and is the most effective leading indicator available. Consequently, the committee has proposed a sequence of efforts designed to construct a robust exposure component of the envisioned surveillance system. The report calls for an immediate collaborative effort of federal agencies to initiate the development of a comprehensive approach for exposure surveillance that builds and updates a database of risks and exposures to predict and locate work-related acute and chronic health conditions for prevention (Recommendation H).

Emphasizing that the overall system relies, at all levels and in all circumstances, on an adequately educated and trained professional workforce, the report also calls for a collaborative federal effort to promote and support education and training of the surveillance workforce. The responsible agencies need to identify core competencies required for OSH surveillance (e.g., epidemiology, biomedical informatics, and biostatistics) and engage educational institutions to establish or modify training programs accordingly (Recommendation P). Steps toward building this trained workforce are proposed that can begin immediately.

## Improve Data Collection

The committee began its effort by adopting the Centers for Disease Control and Prevention's definition of surveillance, which starts with the collection and processing of relevant data closely linked to analysis and interpretation that can guide policy and interventions. The committee accordingly focused on the sources and quality of the inputs to the smarter system. Throughout the report, the issue of undercounting of occupational injuries and illnesses is highlighted from two perspectives: cases that are *out of scope* or cases that *are simply unreported*. Failure to count occupational injuries that are out of scope is a recognized consequence of surveys or other assessments that do not capture data on some segments of the working population. For example, the Survey of Occupational Injuries and Illnesses (SOII) does not cover or capture injuries to workers who are self-employed (e.g., independent contractors) or who work on small farms. When the 1987 NRC report called specific attention to both perspectives, the Bureau of Labor Statistics (BLS) responded by addressing that which was within its control and now has ongoing efforts to understand the reasons for underreporting (BLS, 2017). The underreporting problem, however, is complex and multifactorial (Azaroff et al., 2002). Additionally, there are limitations in the identification and reporting of chronic diseases associated with work. Further, many of the systems that collect information on injuries and illnesses do not collect occupationally-related data.

One of the major inputs to OSH surveillance is through the SOII and the report discusses needed enhancements to the SOII, including those that would better inform surveillance and related public health actions for underserved populations. Injury and illness recording, as defined for the SOII, can be improved by better characterization of work-related injuries and illnesses in a manner that enhances usefulness at the worksite as well as at national and state levels (Recommendation A). Developing ways to incorporate information on race and ethnicity as well as employment arrangements will allow for identification of vulnerable worker populations and risks that may be associated with different types of employment arrangements. Substantial progress toward this recommendation can be achieved in the near term while some parts will require new methods and resources.

The committee supports the BLS plan to implement a Household Survey of Occupational Injuries and Illnesses (HSOII) as it will fill in data gaps for populations of workers who are missing from employer-based injury reporting and will provide worker input (Recommendation D). Already being piloted, this survey would fill in important recognized gaps in the SOII coverage. Another largely untapped resource for injury surveillance data is the workers' compensation system and the report promotes the expanded use of workers' compensation data for occupational injury and illness surveillance (Recommendation F).

Work-related disease information has been almost absent from all efforts at occupational health surveillance. This absence was noted in 1987 and remains today. The committee considers this a priority component of data collection and offers several recommendations to attend to this need. These deal with occupational disease monitoring (Recommendation B), the specification of industry and occupation as core variables in all federal health surveys (Recommendation G), the enhanced assessment of self-reported health through the National Center for Health Statistics or an expanded HSOII (one component of Recommendation D), and the development of a comprehensive approach for exposure surveillance (Recommendations D and H). The latter recommendation addresses the unparalleled opportunity to gain information on the distribution of exposure-related factors in a manner modeled on the highly successful experience in the European Union, which has over 25 years of experience in such efforts. The committee acknowledges that full implementation of these five recommendations will require careful planning and a long-term effort. But there are near-term steps that move toward the end goals that warrant immediate attention.

## Expand Biomedical Informatics Use and Capabilities

The effectiveness of the overall system of systems will necessarily depend on utilizing the evolving resources and methods of biomedical informatics. Developments occur so rapidly in this area that a lack of experienced, engaged personnel leads to lost opportunities and compromised system effectiveness. The committee considered several aspects of the informatics need but it would be hard to overemphasize how critically important it is for NIOSH to attract adequate informatics personnel and resources (Recommendation J), even while acknowledging the difficulties in recruiting and maintaining informatics experts in the public sector. Chapter 7 offers several specific examples of how informatics capacity can be leveraged to enable NIOSH and other OSH agencies both to use advanced computational and analytical tools and to monitor advances in information for the most effective OSH surveillance (Recommendations K, L, and M).

Achieving the goals of these three recommendations is expected to take some time but it is essential that the initial steps not be delayed.

For occupational health surveillance, a forward-looking aspect of data collection and processing concerns how best to remove the barrier to recording and interpreting occupational information in medical records and in population surveys of all types. The report accordingly recommends that NIOSH, with an evolving biomedical informatics capacity, lead efforts to establish data standards and software tools for coding and using occupational data in electronic health records. As these records increasingly become the standard for practice, there is an opportunity to make substantial long-lasting progress to eliminate barriers to linking occupation and disease wherever necessary (Recommendation L). The committee also calls for the creation of a cross-agency effort to develop and evaluate state-of-the-art computational and analytical tools for processing free-text data found in occupational safety and health records of all types (Recommendation M). The benefits that accrue from action on these recommendations will prove invaluable in several arenas in addition to surveillance.

### Strengthen Data Analysis and Information Dissemination for Prevention

Successful collection and processing of surveillance data alone does not make a successful surveillance system. The system also requires thoughtful analysis, careful interpretation, and then dissemination to all those who can use the results to engage in public health action for prevention. The committee accordingly also examined needs at this stage of surveillance and is optimistic that there are opportunities immediately available, as well as ones that one can realistically imagine, that can support the pursuit of a smarter OSH surveillance system.

Attention to analysis and interpretation is essential when calling on partners to provide new or more data, even if that requirement is facilitated through electronic reporting. It is accordingly essential that a program that provides for better reporting, such as the OSHA electronic reporting initiative, needs to be accompanied by a robust plan for the analysis, interpretation, and dissemination of the resulting information. The committee thus calls on OSHA and its sister agencies and stakeholders to develop and publicize plans to maximize the utility of their new electronic reporting initiative by providing means and methods for ongoing analysis and dissemination of these data with special attention to serving individual employer needs, while simultaneously minimizing duplication of reporting by employers (Recommendation E).

State and federal workers' compensation systems, though offering great potential to contribute important insights into the causes of occupational injury and the effectiveness of prevention programs, are constrained by differences in system architecture and coding approaches when compared to other sources of injury data. The report accordingly calls on NIOSH, with assistance from OSHA, to promote the expanded use of workers' compensation data for occupational injury and illness surveillance and to begin to consider the development of surveillance for outcomes or consequences of injury and illness outcomes (Recommendation F).

Dissemination of surveillance findings and analyses in forms and substance so these can be used to inform and evaluate prevention is equal in importance to all that has been discussed thus far. The committee has presented the case for making regular reports to the nation that publicize the overall burden of occupational injury and disease in terms of the burdens on health, the economy, and society so these burdens can be better characterized (Recommendation I and Q). The committee also recognizes that NIOSH, OSHA, and BLS are practiced in dissemination, and while the committee provides ideas throughout the report on how these efforts could be enhanced to better serve the prevention goal, there are two areas that rose to the level of formal recommendations. The first is engaging with the scientific community by working with the National Library of Medicine to facilitate easy discovery of important connections between work and disease or injury in published research (Recommendation K). Importantly, the committee also recommends the creation of a smoothly operating alert mechanism that receives, enhances, and rapidly publicizes to those who need to act the signals of emerging occupational health problems either as new associations of work and illness or injury or of old associations found in new settings (Recommendation N). Appropriate and timely attention to surveillance findings, routine or new, is essential for prevention and thus requires that a smooth and centralized mechanism or office be established for timely ongoing dissemination of information to all relevant actors (Recommendation O).

## CONCLUDING REMARKS

Worker safety and health is of paramount importance to thriving workers and workplaces, and accordingly to society as a whole. Ensuring and improving worker safety and health is a commitment taken seriously, and diligently acted upon, by numerous federal, state, and local agencies; workers and worker organizations; employers and employer organizations; and many others. More can be done to inform and improve these efforts through strengthening OSH surveillance in the United States. With the rapid changes in the nature of work in the United States, and with new risks added to those that have always existed, the nation clearly needs a smarter OSH surveillance system of systems for the 21st century. This report provides the evidence and recommendations for a greatly enhanced

OSH surveillance system that is envisioned to be smarter, more dynamic, and more highly coordinated.

## REFERENCES

Azaroff, L. S., C. Levenstein, and D. H. Wegman. 2002. Occupational injury and illness surveillance: Conceptual filters explain underreporting. *American Journal of Public Health* 92(9):1421-1429.

BLS (Bureau of Labor Statistics). 2017. *Research on the completeness of the injury and illness counts from the Survey of Occupational Injuries and Illnesses*. Available online at https://www.bls.gov/iif/undercount.htm (accessed August 21, 2017).

NRC (National Research Council) 1987. *Counting Injuries and Illnesses in the Workplace: Proposals for a Better System.* Washington, DC: National Academy Press.

# Appendix A

# Recommendations

**Recommendation A (Chapter 4): BLS and OSHA should collaborate to enhance injury and illness recording and the SOII to achieve more complete, accurate, and robust information on the extent, distribution, and characteristics of work-related injuries and illnesses and affected workers for use at the worksite and at national and state levels.** As part of this effort, BLS should routinely collect detailed case and demographic data for injuries and illnesses resulting in job transfer or restricted-duty work. Furthermore, OSHA should amend its injury and illness recording requirements to collect information on race and ethnicity as well as on employment arrangement to identify vulnerable worker populations and risks that may be associated with the changing nature of work.

In the near term:

- OSHA should make type of employment arrangement (e.g., traditional, independent contractor, temporary agency worker, and on-call worker) and race and ethnicity mandatory data elements on the OSHA Form 301, and BLS should incorporate this information into the SOII case and demographic data. OSHA should collaborate with BLS in determining the best approach to collecting this information (e.g., what questions should be included on Form 301).
- BLS should routinely collect detailed case and demographic data for injuries and illnesses resulting in job transfer or restricted duty as well as those resulting in days away from work.

• BLS should implement methods to aggregate SOII data over time to generate more robust and detailed state-level estimates.

• OSHA and BLS should collaborate to enhance recordkeeping training for employers and BLS should evaluate approaches for providing initial information and ongoing feedback to data recorders in establishments enlisted to participate in the SOII both to improve the data quality and to promote employer use of data for prevention.

In the longer term:

• BLS and OSHA should collaborate to determine the best way to collect injury and illness data across multiple employers working at single sites, and across enterprises with multiple establishments.

• BLS should assess the feasibility and usefulness of extending collection of case and demographic data to all reported cases as automated approaches to coding SOII narratives are improved. Options for collecting such data should be evaluated in light of information that will be made available through the OSHA electronic reporting initiative.

**Recommendation B (Chapter 4): NIOSH, working with the state occupational safety and health surveillance programs and across divisions within the agency, should develop a methodology and coordinated system for surveillance of both fatal and nonfatal occupational disease using multiple data sources.** The data should be analyzed, interpreted, and presented regularly in a comprehensive public report. The data sources to be considered should include reporting by audiometric providers, disease registries (such as cancer and chronic renal failure), hospitals, laboratories, physicians, poison control centers, and health surveys as well as appropriate exposure databases. It is important that illness data collected by the states and NIOSH be analyzed and released in a timely manner. The data should be released in conjunction with BLS illness data in a manner that does not delay data released by BLS.

In the near term:

• NIOSH should combine information from the existing focused occupational disease surveillance systems (e.g., ABLES, pesticide illness, silicosis surveillance, and NORMS) and mesothelioma from cancer registries and other relevant occupational health indicators to provide a more comprehensive annual report on the extent of occupational illness morbidity and mortality that can be released in conjunction with information from the SOII. Methods for extrapolating from available data to generate national estimates should be explored.

• To enhance surveillance of occupational lung disease, NIOSH should require all B readers to report all chest radiographs interpreted to be positive for all types of pneumoconiosis.

• Increased collaboration between NIOSH and CDC infectious disease surveillance programs, with improved collection of occupational information, will be important to improve documentation of endemic and epidemic infectious disease related to work.

In the longer term:

• Gaps identified in the occupational illness surveillance system will need to be addressed through future developments that may involve

  o New or modified state regulations, requiring close coordination with the states, many of which have already promulgated reporting regulations.

  o Inference of occupational disease burden and trends that result from enhanced exposure assessment (Recommendation H, see Chapter 6).

  o Updating the list of occupational sentinel health events, establishing additional criteria for establishing a link between workplace exposures and common diseases.

• Action on recommendations that address the inclusion of occupational information in medical records (see Recommendation J, see Chapter 7), federal health surveys and public health surveillance systems (Recommendation M, see Chapter 7), and automated coding of the industry and occupation information (see Recommendation L, see Chapter 7) will be important for ensuring the optimal implementation over time of this recommendation.

**Recommendation C (Chapter 4): NIOSH should lead a collaborative effort with BLS, OSHA, the states, and other relevant federal agencies to establish and strengthen state-based OSH surveillance programs.** This should be carried out as part of a national coordinated effort to monitor priority conditions, hazards, and exposures; to identify emerging workplace risks; and to facilitate prevention programs that address these concerns. Furthermore, this should be carried out with the full support of and assistance from other parts of HHS-CDC.

In the near term:

• *OSH Agency Collaboration Within States:* NIOSH, BLS, and OSHA should actively encourage and promote collaboration among their programs in the states to reflect the national commitment to interagency effectiveness for OSH surveillance and leverage surveillance and prevention expertise across agencies. This should include sharing data and taking advantage of unique state-level data sets and case-based surveillance capacities to identify and respond to emerging occupational safety and health hazards and conditions.

- *Public Health Agency Collaboration Within States:* NIOSH and other CDC centers that support state-based surveillance and prevention activities should promote collaboration among their state-level programs to monitor and address public health problems of shared concern, such as violence, asthma, infectious disease, traffic safety, and health inequities among vulnerable population groups.
- NIOSH should also
    - o Explore and implement, as appropriate, alternative approaches to funding ongoing surveillance in the states as applied public health programs rather than research programs.
    - o Foster increased coordination and communication between its intramural and extramural surveillance programs.
    - o Encourage NIOSH-funded Education Research and Training Centers and Agricultural Health and Safety Centers to provide technical and research support to state surveillance programs in their regions as part of their required outreach and education core activity.

**Recommendation D (Chapter 6): BLS should place priority on implementing its plan for a household survey of nonfatal occupational injury and illnesses (HSOII).** With the assistance of NIOSH and Centers for Disease Control and Prevention (CDC), BLS should also expand this effort to include a periodic nationwide household survey to identify and track reports of occupational exposures and should determine how best to identify and track chronic work-related illnesses.

In the near term:

- BLS should survey occupational injuries and acute illnesses (as in SOII) in a nationally representative sample of the entire working population including those who are self-employed or engaged in temporary contract work.

In the longer term:

- To address the inadequacies of current surveillance tools, BLS should
    - o Seek assistance from NIOSH to enhance the HSOII survey scope by assessing occupational exposures and risks in a manner like that used in the Eurofound Working Conditions Survey.

o Questions should be included to capture exposure determinants and work characteristics with sufficient details on industry, occupation, work organizational characteristics, and working relationships in a way that supports the development of a flexible job exposure matrix and supports integration of newly available or ancillary data.

o Seek assistance from NCHS and NIOSH to address currently inadequate information on chronic disease and work by determining whether self-report of illnesses and chronic conditions are best tracked by inserting occupational information into the NHIS or inserting chronic illness questions into the HSOII. Part of this consideration should include the determination of whether a sample of retirees and those not working due to disability should be part of the HSOII.

- BLS should prepare and implement a specific plan for routine analysis, interpretation, and preparation of a report on the findings from the HSOII along with a plan for dissemination and appropriate database access by researchers and the public.

**Recommendation E (Chapter 6): OSHA, in conjunction with BLS, NIOSH, state agencies, and other stakeholders, should develop plans to maximize the effectiveness and utility of OSHA's new electronic reporting initiative for surveillance.** These should include plans to provide ongoing analysis and dissemination of these data and to minimize duplication of reporting by employers.

In the near term:

- To avoid duplicate reporting, OSHA and BLS should integrate data-collection efforts so that employers selected in the annual BLS sample for SOII but reporting electronically to OSHA need not make separate reports to BLS. This will require that a unified reporting form include requiring race and ethnicity in submitted case reports.
- OSHA should provide timely and automatic feedback to employers that provides comparative information specific to the employer and others in that industry.
- OSHA should develop a publicly available and easily searchable injury and illness database based on the electronic reports.

In the longer term:

- OSHA and NIOSH should work with stakeholders to develop software and other tools and materials that facilitate further establishment-level analysis of injury data with specific attention to enabling effective use by employers as well as others to identify hazards and job-specific issues for prevention. With experience from participants in this electronic reporting, OSHA should explore feasibility to expand electronic reporting to all employers required to maintain OSHA logs.

**Recommendation F (Chapter 6): NIOSH with assistance from OSHA should explore and promote the expanded use of workers' compensation data for occupational injury and illness surveillance and the development of surveillance for consequences of injury and illness outcomes, including return to work and disability.**

In the near term:

- NIOSH should organize an advisory group of workers' compensation data experts to advise both the NIOSH Center for Workers' Compensation Studies and interested states concerning their use of workers' compensation data for surveillance and research.
- NIOSH should encourage states to expand the use of workers' compensation information beyond the Council of States and Territorial Epidemiologists (CSTE) occupational health indicators. Specifically, the agency should work through the state surveillance cooperative agreements to develop and enhance use of workers' compensation data for state-based occupational injury and illness surveillance and prevention activities.

In the longer term:

- NIOSH and OSHA should collaborate with states to pursue the development of surveillance systems that capture cost of work-related injury and illness, measure work-related disability and return-to-work outcomes, and assess the adequacy of benefits administered through workers' compensation insurance programs.

**Recommendation G (Chapter 6): HHS should designate industry and occupation as core demographic variables collected in federal health surveys, as well as in other relevant public health surveillance systems, and foster collaboration between NIOSH and other CDC centers in maximizing the surveillance benefits of including industry and occupation in these surveys and surveillance systems.**

In the near term:

- HHS should re-establish industry and occupation as core demographic variables in all federal health surveys.
- CDC surveillance programs, as they proceed with their state part-

ners to streamline and harmonize data across systems, should work with NIOSH to identify appropriate processes for collecting and coding occupational and industry data.

- NIOSH with assistance from CDC should explore and prioritize public health surveys that can be used to enhance occupational health surveillance objectives by collecting relevant occupational information.

In the longer term:

- To promote proper analysis of surveillance data NIOSH should develop methods and training materials on approaches to basic as well as new and creative use of occupation and industry data and on the selection and use of appropriate labor force denominators.

**Recommendation H (Chapter 6): NIOSH, in consultation with OSHA, should place priority on developing a comprehensive approach for exposure surveillance.** The objective should be to build systematically a comprehensive and continuously updated database of risks and exposures that provides the basis for estimating work-related acute and chronic health conditions for prevention.

In the near term:

- NIOSH should fully exploit the existing OSHA exposure databases by cleaning and integrating all available data sources to make them useful for surveillance purposes, taking proper account of the database limitations.

As an intermediate goal:

- NIOSH, in collaboration with OSHA and other agencies as appropriate, should construct an integrated exposure database to include the multiple sources of exposure measurement data already available, specifically MSHA's MSIS, Department of Energy and Nuclear Regulatory Commission personal exposure data, and relevant data from others conducting research with federal funds.

In the longer term:

- NIOSH should link the integrated exposure database with the comprehensive survey data obtained in the recommended expanded HSOII (Recommendation D) and new data from any characterization of exposures from targeted industry-specific assessments.

• NIOSH and OSHA should explore the feasibility of receiving employer-mandated exposure sample results after considering the reliability and quality of those measurements. The agencies should work with stakeholders to develop software and other tools and to facilitate establishment-level analysis of exposure data along with benchmarking.

**Recommendation I (Chapter 7): NIOSH should coordinate with OSHA, BLS, and other relevant agencies to measure and report, on a regular basis, the economic and health burdens of occupational injury and disease at the national level.** This report should also attempt to address the contribution of implemented interventions in reducing these burdens. The advantages of a regular, standard report on national economic burden of occupational injury and disease include:

• focusing attention on the significant burden that already exists,
• measuring progress over time in reducing those burdens and improving worker safety and health,
• improving the allocation of existing resources to improve health outcomes, and
• establishing priorities.

Research, such as to establish the fraction of disease and injuries attributable to occupational exposures, will be necessary to continually improve the quality of burden estimates that can be produced.

**Recommendation J (Chapter 7): NIOSH should build and maintain a robust internal capacity in biomedical informatics applied to OSH surveillance.**

In the near term:

• Assess the need within the agency for expertise in biomedical informatics in the context of current and future demand, recognizing that it will be important to train informatics talent in OSH surveillance and then to work to retain talented individuals who develop knowledge at the intersection of the informatics discipline and OSH applications;
• Create an organizational strategy for deploying and making optimal use of expertise in biomedical informatics to support the planning and conduct of OSH surveillance;
• Develop a plan for hiring, including consideration of steps such as reaching out to academic programs, advertising in different venues, and offering internships; and
• Develop a plan for retention, including opportunities for continuing education.

**Recommendation K (Chapter 7): NIOSH should work with the National Library of Medicine to incorporate core OSH surveillance terminologies, including those for industry and occupation, into the Unified Medical Language System (UMLS).** The creation and maintenance of mappings among OSH terminologies and between OSH terminologies and other relevant terminologies already included in the UMLS should be considered.

In the near term:

- Establish an inventory of relevant OSH terminologies;
- Develop use cases that benefit from the existence of mappings across OSH terminologies; and
- Prioritize terminologies in terms of the value that accrues from incorporating them into the UMLS.

In the longer term:

- Incorporate highest-priority OSH terminologies into the UMLS.

**Recommendation L (Chapter 7): NIOSH should lead efforts to establish data standards and software tools for coding and using occupational data in electronic health records.** These efforts should be coordinated with the Office of the National Coordinator for Health Information Technology (ONC) to support the establishment of a rule requiring collection and effective use of OSH data in the electronic health record.

In the near term:

- Develop a consensus within the OSH surveillance community regarding the preferred terminologies and tools for extracting data on industry and occupation from the EHR;
- Engage with ONC to communicate this consensus to other stakeholders and to establish a broader consensus among all stakeholders regarding an acceptable strategy; and
- Support ONC in the process of establishing a rule to require the capture of industry and occupation in the EHR.

In the longer term:

- Work with the occupational medicine and general medicine community to develop models and tools for using occupational data in electronic health records for clinical care and for serving the prevention needs of the clinical population.

**Recommendation M (Chapter 7): NIOSH and BLS, working with other relevant agencies, academic centers, and other stakeholders should coordinate and consolidate, where possible, efforts to develop and evaluate state-of-the-art computational and analytical tools for processing free-text data found in OSH surveillance records of all types.** This coordination should enable rapid innovation and implementation, into OSH surveillance practice, of successful "autocoding" methods for different data sources.

In the near term:

- Conduct an inventory of activities and key stakeholders and
- Support knowledge exchange activities (symposia, competitions).

In the longer term:

- Develop open data sets that can be used to consistently evaluate methods for extracting OSH data from free text.

**Recommendation N (Chapter 7): To identify emerging and serious OSH injuries, illnesses, and exposures in a timely fashion, NIOSH (in coordination with OSHA, BLS, and the states) should develop and implement a plan for routine, coordinated, rapid analysis of case-level OSH data collected by different surveillance systems, followed by the timely sharing of the findings.**

In the near term:

- Develop analytical objectives, identifying the outcomes that would benefit from routine, rapid analysis and continuous monitoring across OSH surveillance systems; and
- Review technical and legal strategies for conducting analyses, including novel analytical methods and strategies for distributed analysis and ongoing analysis as the data evolve over time.

In the longer term:

- Implement routine processes for rapid data analysis, including protocols to guide the interpretation of alerts.

**Recommendation O (Chapter 7): To promote and facilitate the use of surveillance information for prevention, and to present more comprehensive information on the extent, distribution, and characteristics of OSH injuries, illnesses, and exposures, NIOSH (in coordination with and input from OSHA, BLS, and the states) should establish a coordinated strategy and mechanism for timely dissemination of surveillance information.**

In the near term:

- Clarify target populations for different types of surveillance information (e.g., rapid alerts, trends, etc.);
- Establish a plan for accessing, integrating, and disseminating information from different surveillance sources; and
- Develop policies and criteria to address individuals' and employers' privacy and confidentiality considerations through a process that provides for stakeholder input and includes privacy experts in the development of these policies and in the design of surveillance systems.

In the longer term:

- Implement a coordinated information dissemination strategy, making use of different technologies as appropriate to communicate information to those who need it to take action for prevention.

**Recommendation P (Chapter 7): NIOSH, OSHA, and BLS should work together to encourage education and training of the surveillance workforce in disciplines necessary for developing and using surveillance systems, including epidemiology, biomedical informatics, and biostatistics.**

In the near term:

- Identify the core competencies required for OSH surveillance and promote the science of surveillance;
- Review the curricula of existing surveillance courses;
- Collaborate with educational organizations to establish or modify training programs accordingly; and
- Require surveillance courses in all funded training programs, especially in the Education and Research Center and Program Project training grants.

In the longer term:

- Contribute to development of surveillance courses and conferences that provide training in surveillance methods.

**Recommendation Q (meta-recommendation, Chapter 8): The Secretary of Health and Human Services, with the support of the Secretary of Labor, should direct NIOSH to form and lead a coordinating entity in partnership with OSHA, BLS, and other relevant agencies.** The coordinating entity should:

- develop and regularly update a national occupational safety and health surveillance strategic plan that is based on well-articulated objectives;
- coordinate the design and evaluation of an evolving national system of systems for OSH surveillance and for the dissemination of surveillance information provided by these systems;
- publish a report on progress toward the strategic plan's implementation at least every 5 years, documenting advances toward achieving a 21st-century smarter occupational safety and health (OSH) surveillance system; and
- engage partners, including other federal health statistics agencies, state agencies with OSH responsibilities, and stakeholders.

# Appendix B

# Committee Biosketches

**Edward H. Shortliffe** is a professor of biomedical informatics at Arizona State University. He is also an adjunct professor of biomedical informatics at Columbia University, and an adjunct professor of health policy and research (health informatics) at Weill Cornell Medical College. His research interests include the broad range of issues related to integrated decision-support systems, their effective implementation, and the role of the Internet in health care. Previously, he served as the president and chief executive officer of the American Medical Informatics Association. He has also served on the faculty of the University of Texas Health Science Center and the University of Arizona College of Medicine. Before that, he was the Rolf A. Scholdager professor and chair of the Department of Biomedical Informatics at Columbia University College of Physicians and Surgeons and a professor of medicine and of computer science at Stanford University. He is a master of the American College of Physicians and editor-in-chief of the *Journal of Biomedical Informatics*. Dr. Shortliffe is a fellow of the American College of Medical Informatics and the American Association for Artificial Intelligence and an elected member of the American Society for Clinical Investigation and the Association of American Physicians. Dr. Shortliffe is an elected member of the National Academy of Medicine. Dr. Shortliffe received his A.B. in applied mathematics from Harvard College and an M.D. and a Ph.D. in medical information sciences from Stanford University.

**David K. Bonauto** is research director for the Washington State Department of Labor and Industries' Safety and Health Assessment and Research for Prevention (SHARP) program. He is also a clinical associate professor in the Department of Environmental and Occupational Health Sciences at the University of Washington. Dr. Bonauto leads Washington State's occupational injury and illness surveillance program. Dr. Bonauto and SHARP have developed novel approaches to using state workers' compensation data for public health surveillance and research, linking state administrative databases to develop more accurate estimates of the working population at risk, and using population based surveys to assess worker-reported work-related injury and illness rates. Dr. Bonauto has served on the NIOSH Board of Scientific Counselors and is a member of the CSTE Occupational Health Workgroup, a collaboration of state-based occupational injury and illness surveillance programs. Dr. Bonauto received his M.D. from Columbia University, M.P.H. from the University of Washington School of Public Health and Community Medicine, and B.A. from Bowdoin College.

**David L. Buckeridge** is a professor in the department of epidemiology, biostatistics, and occupational health at McGill University in Montreal, Canada. His research is on informatics of public health surveillance, with particular interest in the development and evaluation of methods for surveillance systems that use clinical and administrative data. His previous and ongoing work includes the development of statistical methods for outbreak detection and the use of simulation modeling to evaluate surveillance systems. He also holds a Canadian Institutes of Health Research Chair in e-Health Interventions. He is a Fellow of the Royal College of Physicians and Surgeons of Canada with specialty training in Public Health and Preventive Medicine. Dr. Buckeridge received his M.D. in community medicine from Queen's University, Ph.D. in biomedical informatics from Stanford University, and M.Sc. in epidemiology from the University of Toronto.

**Steven B. Cohen** is the vice president of RTI International's Division for Statistical and Data Sciences. He has been working in the fields of biostatistics, survey design, sampling, survey methods and health services research for more than 35 years. He has expertise in management, health services research, health policy, biostatistics, sampling theory, modeling, complex survey design, multivariate analysis, demographic techniques, epidemiological techniques, categorical data analysis, and applied statistical methods. Prior to coming to RTI, he was director of the Center for Financing, Access and Cost Trends at the Agency for Healthcare Research and Quality. He is co-author of the text, Methodological Issues for Health Care Surveys. He has served as an associate professor at the Johns Hopkins University and at the George Washington University. He is also a Fellow of the American Statistical Association and an elected member of the International Statistical Institute. Dr. Cohen received his Ph.D. (1978) in biostatistics from the University of North Carolina, M.S. (1976) in biostatistics from the University of North Carolina, and B.A. (1973) in mathematics and history from the City University of New York, Brooklyn College.

**Letitia K. Davis** is director of the Occupational Health Surveillance Program in the Massachusetts Department of Public Health where she has worked for over 30 years to develop and implement state-based surveillance systems for work-related illnesses and injuries. She has overseen the formation of a health care provider reporting system for occupational disease, the Massachusetts Occupational Lead Registry, a comprehensive surveillance system for fatal occupational injuries, the Massachusetts Sharps Injury Surveillance System, a surveillance system for work-related asthma, and a model surveillance system for work-related injuries to young workers. Additionally, she has overseen implementation of case-based surveillance and follow-up of work-related amputations, burns and acute chemical poisonings. She has conducted numerous surveillance research studies exploring use of a wide range of public health data sources for occupational health surveillance, including a recent study exploring the feasibility of multisource surveillance for work-related amputations and carpal tunnel syndrome. She has a particular interest in addressing the occupational health and safety concerns of vulnerable workers and has recently completed a project incorporating occupational information in the electronic record systems of community health centers to improve documentation of occupational health needs of underserved worker populations. From 1998-2015, Dr. Davis was also a lead consultant in occupational health to the Council of State and Territorial Epidemiologists and has played a leadership role nationally in the effort to integrate occupational health into public health practice at the state level. She is a past member of the Board of Scientific Counselors of the National Institute for Occupational Safety and is a current member of the National Advisory Committee on Construction Safety and Health. She has also served on a number of Institute of Medicine committees, including most recently a committee addressing incorporation of occupational information in electronic health records. Dr. Davis serves as adjunct faculty of the Department of Work Environment at the University of Massachusetts at Lowell and an instructor at the Harvard School of Public Health. Dr. Davis received her doctorate in occupational health from the Harvard School of Public Health in 1983.

**Gerald F. Kominski** is a professor of health policy and management and director of the UCLA Center for Health Policy Research. His research focuses on evaluating the costs and cost-effectiveness of health care programs and technologies, with special emphasis on public insurance programs, including Medicare, Medicaid, and workers' compensation. He is also working extensively on evaluating the expected and actual impacts of health care reform and has co-led the development of a microsimulation model (CalSIM) for forecasting eligibility, enrollment, and expenditures under health reform. From 2003-2009, he served as vice chair for the Cost Impact Analysis Team of the California Health Benefits Review Program, which conducts legislative analyses for the California legislature of proposals to expand mandated insurance benefits. Dr. Kominski received his Ph.D. in public policy analysis from the University of Pennsylvania Wharton School in 1985, and his A.B. from the University of Chicago in 1978.

**Scott A. Mugno** is the vice president for safety, sustainability, and vehicle maintenance at FedEx Ground. He was previously the managing director for FedEx Express Corporate Safety, Health, and Fire Protection where he developed, promoted, and facilitated the safety and health program and culture for all nonflight FedEx Express domestic operations. His department also provided technical support to the FedEx Express international operations and other FedEx operating companies. Mr. Mugno has been in the environmental, health, safety, or transportation arenas for over 25 years. He joined FedEx Express as a senior attorney in the Legal and Regulatory Affairs Department. Prior to FedEx, Mr. Mugno was division counsel at Westinghouse Electric Corporation's Waste Isolation Division and deputy staff judge advocate for the Eastern Region U.S. Army Military Traffic Management command. He has held other legal positions in the Army JAG Corps and in private-practice law firms. Mr. Mugno most recently served on the IOM Committee on Health Threats Resilience and Workforce Resilience. He regularly represents FedEx at various trade and safety association and committee meetings and is a frequent speaker before those and other groups. Mr. Mugno received his J.D. from Washburn University School of Law.

**Kenneth D. Rosenman** is a professor of medicine and chief of the Division of Occupational and Environmental Medicine at Michigan State University (MSU). For the last 28 years under the direction of Dr. Rosenman, the occupational and environmental health team at Michigan State University has worked closely with Michigan Occupational Safety and Health Administration and the Michigan Department of Health and Human Services to administer Michigan's occupational injury and illness surveillance program. He has an active research program in occupational and environmental disease with particular interest in the methodology for tracking occupational conditions and the etiology of pulmonary disease. He has published 205 articles on occupational and environmental disease. Prior to joining the faculty at MSU he taught in the University of Massachusetts School of Public Health and was director of occupational and environmental health services at the New Jersey Department of Health. He is board-certified in internal medicine and in occupational and environmental medicine. He is a Fellow of the American College of Epidemiology, the American College of Occupational and Environmental Medicine, the American College of Preventive Medicine, and the Collegium Ramazzini. Dr. Rosenman received his M.D. from New York Medical College.

**Noah S. Seixas** is professor of exposure sciences at the University of Washington School of Public Health. Dr. Seixas has taught at the University of Washington since 1993 and currently serves as chief editor of the *Annals of Occupational Hygiene*. His research interests involve exposure assessment and control with a range of applications including welding fume exposure and control, noise in the construction industry, and injury risk prevention, and safety and health management addressing occupational health disparities. Dr. Seixas received an M.S. in industrial hygiene at Harvard School of Public Health in 1982 and a Ph.D. in industrial health from the University of Michigan in 1990.

**Margaret (Peg) M. Seminario** is the director of safety and health for the AFL-CIO. She has worked for the AFL-CIO since 1977, and since 1990 has been responsible for directing the AFL-CIO's program on safety and health. She has worked extensively on a wide range of regulatory and legislative initiatives at the federal and state levels and coordinated the labor movement's campaigns on Right to Know, ergonomics, and other key job safety issues. She has participated in dozens of rulemakings on important OSHA standards including rules to protect workers from asbestos, lead, hazardous chemicals, and safety hazards like confined spaces. She was one of the leaders in labor's efforts to enact the 9/11 Health and Compensation Act to provide health care and compensation to responders sick from exposures at the World Trade Center. Ms. Seminario has served on numerous federal agency and scientific advisory committees and participated in international safety and health work through the ILO, the OECD and international trade union organizations. Ms. Seminario previously served on the 1987 NRC Panel on Occupational Safety and Health Statistics. She holds an M.S. degree in industrial hygiene from the Harvard School of Public Health and a B.A. in biological sciences from Wellesley College.

**Glenn M. Shor** is the manager of the California Census of Fatal Occupational Injuries Program and the research and policy advisor to the director of the California Department of Industrial Relations. Since 2012, he has provided oversight and coordination of research programs in workers' compensation and occupational safety and health; assistance to Cal/OSHA on program statistics, inspection targeting, and resource allocation; oversees management and policy development for California's Workers' Compensation Information System, and is the Department of Industrial Relations representative on the Workers' Compensation Fraud Assessment Panel.

He previously served as special assistant to the Assistant Secretary of the Department of Labor's Occupational Safety and Health Administration. In this role, he collaborated with medical staff on design and implementation of web page oriented to clinician understanding of OSHA and occupational health and safety practice. He was also involved in developing regulatory proposals on injury and illness prevention program standards, and served as agency coordinator and department lead on a joint OSHA/NIOSH/BLS conference on use of workers' compensation data for injury and illness prevention and public health activities. In 2012, the International Association of Industrial Accidents Boards and Commissions awarded him the Samuel Gompers Award for his work on identifying, educating, and promoting injury and illness prevention activities within workers' compensation. Dr. Shor is an elected member of the National Academy of Social Insurance. Dr. Shor received his M.P.P. and Ph.D. in public policy from the University of California at Berkeley, his B.A. in urban studies from the University of Pennsylvania, and studied epidemiology at the University of Minnesota's School of Public Health.

**David H. Wegman** is emeritus professor at the University of Massachusetts Lowell. Dr. Wegman was previously dean of the School of Health and Environment a position he assumed after 16 years as a professor and founding chair of the Department of Work Environment. He also serves as adjunct professor at the Harvard School of Public Health and as vice president and a member of the Board of Directors of the Alpha Foundation for the Improvement of Mine Safety and Health. Dr. Wegman has focused his research on epidemiologic studies across a range of work-related health conditions, including respiratory disease, musculoskeletal disorders, kidney disease, and cancer. He has also written on public health and policy issues concerning occupational hazard and health surveillance, methods of exposure assessment for occupational epidemiologic studies, the development of alternatives to occupational regulation and the use of participatory methods to study occupational health risks. Dr. Wegman served two terms as a member of the NRC Board on Human Systems Integration and currently serves on the IOM Committee on Aerospace Medicine and the Medicine of Extreme Environments. He has served as both member and chair of over 15 NRC and IOM committees, and has been designated as a National Associate of the Academies. Dr. Wegman is a contributing editor for the *American Journal of Industrial Medicine*, and served as a member of the editorial board of the *American Journal of Public Health*, the *Journal of Occupational Health Psychology*, and the *International Journal of Occupational and Environmental Health*. Dr. Wegman received his B.A. from Swarthmore College, and his M.D. and M.Sc. from Harvard University and is board certified in preventive medicine (occupational medicine).

# Appendix C

# Open Session Meeting Agendas

## FIRST MEETING AGENDA

June 15-16, 2016
Keck Center of the National Academies
Room 101

*WEDNESDAY, June 15*

| | |
|---|---|
| 1:00 – 1:15 p.m. | **Welcome and Introductions**<br>Ted Shortliffe, Committee Chair |
| 1:15 – 1:20 p.m. | **Remarks from Academies' Boards Overseeing Study**<br>*Robin Schoen, Director of Board on Agriculture and Natural Resources*<br>*Connie Citro, Director of Committee on National Statistics*<br>*Andy Pope, Director of Board on Health Sciences Policy* |
| 1:20 – 1:30 p.m. | **NAS Study Process and Committee's Statement of Task**<br>*Peggy Yih, Study Director*<br>(10-minute presentation) |

| | |
|---|---|
| 1:30 – 2:00 p.m. | **Panel: Charge to the Committee from the Sponsors**<br>*Margaret Kitt (NIOSH), Kristen Monaco (BLS), and David Michaels (OSHA)*<br>(5-minute remarks from each agency to discuss why they are sponsoring the study and how they hope the report will be useful for them; 15-minute Q&A with committee) |
| 2:00 – 3:00 p.m. | **Overview of BLS's work**<br>*Kristen Monaco, BLS*<br>(45-min presentation, 15-min Q&A with committee) |
| 3:00 – 3:15 p.m. | **Break** |
| 3:15 – 4:15 p.m. | **Overview of NIOSH's work**<br>*Dawn Castillo, Terri Schnorr, Marie Sweeney, NIOSH*<br>(45-min presentation, 15-min Q&A with committee) |
| 4:15 – 5:15 p.m. | **Overview of OSHA's work**<br>David Michaels, OSHA<br>(45-min presentation, 15-min Q&A with committee) |
| 5:15 – 5:25 p.m. | **Public Comments**<br>Please register ahead of time |
| 5:25 – 5:30 p.m. | **Chair's Closing Remarks for Day 1**<br>*Ted Shortliffe, Committee Chair* |
| 5:30 p.m. | **Adjourn Meeting for Day 1** |

*THURSDAY, June 16*

| | |
|---|---|
| 8:30 – 8:45 a.m. | **Welcome and Introductions** |
| 8:45 – 9:15 a.m. | **A Perspective of Workplace Safety and Health Issues**<br>*David G. Sarvadi, Keller and Heckman LLP*<br>(20-min presentation, 10-min Q&A with committee) |
| 9:15 – 9:45 a.m. | **Public Health Surveillance**<br>*Paula Yoon, Centers for Disease Control and Prevention*<br>(20-min presentation, 10-min Q&A with committee) |

| | |
|---|---|
| 9:45 – 10:15 a.m. | **Surveillance Tools**<br>*Edward L. Baker, University of North Carolina*<br>(20-min presentation, 10-min Q&A with committee) |
| 10:15 – 10:30 a.m. | **Break** |
| 10:30 – 11:00 a.m. | **Hazard Surveillance**<br>*Noah Seixas, University of Washington*<br>(20-min presentation, 10-min Q&A with committee) |
| 11:00 – 11:30 a.m. | **Economics and Workers Compensation**<br>*J. Paul Leigh, UC Davis*<br>(20-min presentation, 10-min Q&A with committee) |
| 11:30 – 11:55 a.m. | **Public Comments**<br>Please register ahead of time |
| 11:55 a.m.–<br>12:00 p.m. | **Chair's Closing Remarks**<br>*Ted Shortliffe, Committee Chair* |
| 12:00 p.m. | **Adjourn Open Session** |

## SECOND MEETING AGENDA
**September 21-22, 2016**

**Academies Keck Center**
**500 Fifth Street NW, Washington, DC**
**Room 208**

*WEDNESDAY, September 21*

| | |
|---|---|
| 1:00 – 1:15 p.m. | **Welcome and Introductions**<br>*Ted Shortliffe, Committee Chair* |
| 1:15 – 1:30 p.m. | **NAS Study Process and Committee's Statement of Task**<br>Peggy Yih, Study Director |
| 1:30 – 2:00 p.m. | **Problems in Injury Surveillance and Possible Approaches**<br>*Leslie (Les) Boden, Boston University (confirmed)*<br>(20-min presentation, 10-min Q&A with committee) |

| | |
|---|---|
| 2:00 – 2:30 p.m. | **Social Security Disability Insurance**<br>*Paul O'Leary, Social Security Administration (confirmed)*<br>(20-min presentation, 10-min Q&A with committee) |
| 2:30 – 3:00 p.m. | **Overview of Health and Safety Issues from a Labor Union Perspective**<br>*Eric Frumin, Change to Win (confirmed)*<br>(20-min presentation, 10-min Q&A with committee) |
| 3:00 – 3:15 p.m. | Break |
| 3:15 – 3:45 p.m. | **Alcoa Medical Data for Reducing Workplace Injuries**<br>*Linda Cantley, Yale University (confirmed)*<br>(20-min presentation, 10-min Q&A with committee) |
| 3:45 – 4:15 p.m. | **Construction Industry Surveillance and Challenges with Multi-Employer Worksites**<br>*Garrett Burke, ConstructSecure, Inc. (confirmed)*<br>(20-min presentation, 10-min Q&A with committee) |
| 4:15 – 5:00 p.m. | **Occupational Health Data Systems to Reduce Occupational Injuries and Illnesses at Ford Motor Company**<br>*Gordon Reeve, Ford Motor Company (Retired) (confirmed)*<br>(30-min presentation, 15-min Q&A with committee) |
| 5:00 – 5:30 p.m. | **Sponsor Panel for Follow-up Q&A**<br>*Hilery Simpson, Bureau of Labor Statistics (confirmed)*<br>*Dawn Castillo, Terri Schnorr, & Marie Sweeney, NIOSH (confirmed)*<br>*Dave Schmidt, OSHA (confirmed)* |
| 5:30 p.m. | **Chair's Closing Remarks & Adjourn Meeting for Day 1**<br>*Ted Shortliffe, Committee Chair* |

*THURSDAY, September 22*

| | |
|---|---|
| 8:00 – 8:15 a.m. | **Welcome and Introductions**<br>*Ted Shortliffe, Committee Chair* |
| 8:15 – 8:45 a.m. | **Historical Perspective from a Former Assistant Secretary of Labor for OSHA on Workplace Safety and Health**<br>*John L. Henshaw, Cardno ChemRisk (confirmed)*<br>(20-min presentation, 10-min Q&A with committee) |
| 8:45 – 9:15 a.m. | **Overview of NCHS Data Sets and Work Related Health**<br>*Jennifer Parker, National Center for Health Statistics (confirmed)*<br>(20-min presentation, 10-min Q&A with committee) |
| 9:15 – 9:45 a.m. | **Public Comments**<br>Please register ahead of time |
| 9:45 a.m. | **Chair's Closing Remarks and Adjourn Open Session**<br>*Ted Shortliffe, Committee Chair* |

## THIRD MEETING AGENDA
**November 29, 2016**

**National Academy of Sciences Building**
**2101 Constitution Ave NW, Washington, DC**
Room 125

| | |
|---|---|
| 8:30 – 8:45 a.m. | **Welcome, Introductions, and Goals for the Meeting**<br>*Ted Shortliffe, Committee Chair* |
| 8:45 – 9:45 a.m. | **Finnish System for Occupational Safety and Health Surveillance**<br>*Jorma Rantanen (confirmed)*<br>(40-min presentation, 20-min Q&A with committee) |
| 9:45 – 10:00 a.m. | **Public Comments**<br>Please register ahead of time |
| 10:00 a.m. | **Chair's Closing Remarks & Adjourn Open Session**<br>*Ted Shortliffe, Committee Chair* |

# Appendix D

# Updates on Recommendations from the 1987 National Research Council Report *Counting Injuries and Illnesses in the Workplace: Proposals for a Better System*

| 1987 RECOMMENDATIONS | ACTIONS |
|---|---|
| **Annual Survey** | |
| 1. **The BLS annual survey should be modified to permit the collection of detailed data on severe occupational injuries categorized as injuries resulting in death, hospitalization, or outpatient surgery. This will require:**<br>• Modification of the OSHA 200 and 200S to provide categories for admission to a hospital or for out-patient surgery, regardless of whether at a hospital, clinic, physician's office, or the establishment itself. BLS should convene a working group to develop an appropriate classification and corresponding definitions.<br>• Collection and coding of data from the OSHA 101 (or equivalent) for all fatalities, hospitalizations, and outpatient surgery. | OSHA's revised recordkeeping requirements, including modifications to the forms, became effective on January 1, 2002. The OSHA Form 301 (previously Form 101) includes fields asking whether the injured worker was treated in the emergency room or as in-patient within the hospital.<br><br>The BLS Census of Fatal Occupational Injuries (CFOI), created after the 1987 set of recommendations, collects data on all occupational fatalities using a multisource approach (*see also the response to Recommendation 4, below*). |

| 1987 RECOMMENDATIONS | ACTIONS |
|---|---|
| 2. **The OSHA 200 and 200S should be modified to include specific categories of injuries, such as amputations, burns, cuts, fractures, contusion or bruises, sprains/strains/unspecified pain, and other.** | The Survey of Occupational Injuries and Illnesses now collects, codes, and publishes data on many categories of injuries and illnesses including those listed in Recommendation #2. |
| 3. **The annual survey should continue to collect occupational illness data from the OSHA 200 log with the following revision in the distinction between "acute" and "chronic" and in the categories of illness.**<br>• Acute occupational illnesses should be divided into such categories as skin, respiratory, gastrointestinal, nervous system, musculoskeletal, and other. Illnesses would be recorded in this section if the onset of an illness is less than 14 days after the last exposure identified as the probable cause.<br>• Chronic occupational diseases should be divided into such categories as hearing loss, repetitive trauma disorders, illnesses diagnosed as a result of a medical examination required under the OSHA health standard, and other chronic illness. | OSHA's revised recordkeeping requirements became effective on January 1, 2002. The separate recording criteria that distinguished between injuries and illnesses were removed. OSHA Form 300 has six general categories for each injury and illness recorded: injuries, skin disorders, respiratory conditions, poisoning, hearing loss, and all other illnesses.<br><br>The BLS SOII Case and Demographic product allows for additional detailed illness data to be made available (such as the nature of the illness). Specific distinctions between acute and chronic are not made but estimates are published for the categories listed in this recommendation, based on the Occupational Injury and Illness Coding System. |
| 4. The annual survey should be modified to permit the collection of detailed data on occupational illnesses resulting in hospitalization or death, as recommended above for occupational injuries. This will require:<br>• Modification of the OSHA 200 and 200S to provide a category for hospitalization.<br>• Collection by BLS of the OSHA 101 (or equivalent) for all hospitalization and fatalities. | As a result of the 1987 report, BLS collects detailed case data for Days Away From Work (DAFW). Recently, BLS began collecting information on whether DAFW cases include a visit to the emergency room or result in a hospital stay. Pilot efforts are under way to collect data for Days of Job Transfer or Restriction for select industries. Fatal occupational illnesses are not included in the CFOI. |

| 1987 RECOMMENDATIONS | ACTIONS |
|---|---|
| **5. BLS should undertake studies to assess how well employers understand the guidelines for maintaining the OSHA 200 logs and how uniformly they record entries on the logs. These studies might include, but are not limited to, telephone surveys and small test groups.** | BLS has explored this issue in a variety of ways, including the initial undercount qualitative interviews in the mid-2000s, the four state studies, the Washington state follow-back interviews, and the Westat follow-back study. |
| **6. BLS should select a probability sample of establishments and obtain a copy of the OSHA 200 log from each establishment in the sample. BLS should compare these logs against the summary forms submitted by the establishments in the annual survey sample.** | BLS conducted a quality assurance (QA) study in the mid-2000s and has an ongoing QA study to review logs from subsampled establishments against the submissions to SOII. In general, if an injury or illness is on the log, it is included in the establishment's response to the SOII. |
| **7. If the validation study now being carried out by BLS and OSHA in Massachusetts and Missouri proves to be feasible and useful, in terms of being able to gain access to medical records in the sufficient proportion of the establishments, they should be encouraged to consider extending this approach to a broader sample.** | Subsequent research found that gaining regular access to medical records from across the nation is not feasible. This research included using the methodology from the Massachusetts/Missouri study to evaluate the records of a representative sample of employers that reported data to OSHA through the OSHA Data Initiative. |
| **8. BLS should conduct studies to obtain independent medical information on occupational injuries, such as for outpatient surgery and admissions to hospitals other than for observation, to determine the extent to which these injuries have been recorded on the OSHA 200 logs in the establishments in which the injured workers were employed.** | BLS conducted relevant studies in response to this recommendation and noted that the results indicated that obtaining medical data from multiple sources from across the nation would not be feasible. |

| 1987 RECOMMENDATIONS | ACTIONS |
| --- | --- |
| **9. In order to obtain estimates of coverage, BLS should conduct sample surveys of employees to obtain information as to possible injuries sustained in the workplace. These events should be followed back to determine whether they were, in fact, work-related and whether they had been recorded on the OSHA 200 log. Given the unique research opportunity afforded by the 1987 National Health Interview Survey, we urge BLS to apply this kind of case-by-case follow-back.** | Efforts regarding the 1987 National Health Interview Survey are not known. The household SOII pilot now being planned is a survey of employees, but the sample is specifically not linked to sampled SOII establishments. Differences in results between the household and establishment surveys will be available by various characteristics, such as occupation and industry. |
| 10. BLS should work with state agencies to carry out studies in which complete rosters of occupational fatalities are compiled from death certificates, medical examiner records, workers' compensation claims, and reports to OSHA and matched against the OSHA 200 logs in the establishments in which these workers were employed. | CFOI collects data on all occupational fatalities using a multisource approach. Ongoing CFOI efforts at BLS are exploring the identification of occupational fatalities referenced in public documents, such as print media, and providing rapid turnaround in disseminating this information. |
| 11. Data from the BLS annual survey should be fed back to companies by industry and size classification and posted so that employers and employees can see how their company compares with the appropriate referent group. | BLS includes a section in survey materials on "how your injury and illness data are used." BLS provides data to employers, employees, and other data users in many web-based formats, including the incidence rate calculator, flat data files, tables, charts, publications, and other products. |
| **12. Special studies focusing on specific research topics should be conducted by BLS in conjunction with the annual survey, in which OSHA 101s are collected for specific injuries or from specific industries, as determined by BLS.** | Since 1987 BLS has introduced the SOII Case and Demographic expansion and implemented CFOI, which provide extensive data on specific injuries and industries. Additionally, partnerships with NIOSH have focused on respirator use and workplace violence prevention. |

| 1987 RECOMMENDATIONS | ACTIONS |
|---|---|
| **BLS Supplementary Data System** | |
| **13. The Supplementary Data System should be refocused and modified to gather data in greater detail on all injuries rather than collecting a minimal data set from a large number of states. This would permit detailed analyses for a wide range of specific injuries. Grants should be made on a competitive basis to states that can meet the criteria for data detail and quality specified by BLS. Competition for grants for data analyses should also be open to other organizations or individuals (e.g., university researchers) and should encourage proposals for new areas of analyses.** | The Supplementary Data System was replaced with the SOII Case and Demographics and CFOI. |
| **Use of Other Data Systems** | |
| **14. NIOSH should be designated as the lead agency having the responsibility for the development of a comprehensive occupational disease surveillance system that would include the compilation, analysis, and dissemination to occupational illness data. These data would come from national data bases and state health departments, beginning with data that are already available. As part of this system, NIOSH should support the development of follow-back interventions; should develop standardized methods for the detection of recognized occupational illnesses; and should publish periodic reports summarizing the data on occupational disease from the various sources. To accomplish this, NIOSH should request, and Congress approve, appropriation of additional funds.** | Additional funding has not been provided for a comprehensive system, however NIOSH has been able to work with federal and state partners to collect and explore data on national trends in some areas. NIOSH has developed standardized case definitions for elevated blood levels, pesticide poisonings, work-related asthma, and silicosis. |

NOTE: The examples of actions were provided by staff from NIOSH, OSHA, and BLS in response to a request from the committee. The agency responses were edited to provide some examples of the extensive efforts.

# Appendix E

# OSHA Form 300 and Related Pages

OSHA's Form 300 (Rev. 01/2004)

# Log of Work-Related Injuries and Illnesses

**Attention:** This form contains information relating to employee health and must be used in a manner that protects the confidentiality of employees to the extent possible while the information is being used for occupational safety and health purposes.

Year 20__ __

**U.S. Department of Labor**
**Occupational Safety and Health Administration**

Form approved OMB no. 1218-0176

*You must record information about every work-related death and about every work-related injury or illness that involves loss of consciousness, restricted work activity or job transfer, days away from work, or medical treatment beyond first aid. You must also record significant work-related injuries and illnesses that are diagnosed by a physician or licensed health care professional. You must also record work-related injuries and illnesses that meet any of the specific recording criteria listed in 29 CFR Part 1904.8 through 1904.12. Feel free to use two lines for a single case if you need to. You must complete an Injury and Illness Incident Report (OSHA Form 301) or equivalent form for each injury or illness recorded on this form. If you're not sure whether a case is recordable, call your local OSHA office for help.*

Establishment name ____________________

City ____________________ State __________

| Identify the person | | | Describe the case | | | Classify the case | | | | | | | | | | | |
|---|---|---|---|---|---|---|---|---|---|---|---|---|---|---|---|---|---|
| | | | | | | CHECK ONLY ONE box for each case based on the most serious outcome for that case: | | | | Enter the number of days the injured or ill worker was: | | Check the "Injury" column or choose one type of illness: | | | | | |
| | | | | | | | | Remained at Work | | | | (M) | | | | | |
| (A) Case no. | (B) Employee's name | (C) Job title (e.g., Welder) | (D) Date of injury or onset of illness | (E) Where the event occurred (e.g., Loading dock north end) | (F) Describe injury or illness, parts of body affected, and object/substance that directly injured or made person ill (e.g., Second degree burns on right forearm from acetylene torch) | (G) Death | (H) Days away from work | (I) Job transfer or restriction | (J) Other record-able cases | (K) Away from work | (L) On job transfer or restriction | (1) Injury | (2) Skin disorder | (3) Respiratory condition | (4) Poisoning | (5) Hearing loss | (6) All other illnesses |
| ____ | ____ | ____ | ___/___ month/day | ____ | ____ | ☐ | ☐ | ☐ | ☐ | ___ days | ___ days | ☐ | ☐ | ☐ | ☐ | ☐ | ☐ |
| ____ | ____ | ____ | ___/___ month/day | ____ | ____ | ☐ | ☐ | ☐ | ☐ | ___ days | ___ days | ☐ | ☐ | ☐ | ☐ | ☐ | ☐ |
| ____ | ____ | ____ | ___/___ month/day | ____ | ____ | ☐ | ☐ | ☐ | ☐ | ___ days | ___ days | ☐ | ☐ | ☐ | ☐ | ☐ | ☐ |
| ____ | ____ | ____ | ___/___ month/day | ____ | ____ | ☐ | ☐ | ☐ | ☐ | ___ days | ___ days | ☐ | ☐ | ☐ | ☐ | ☐ | ☐ |
| ____ | ____ | ____ | ___/___ month/day | ____ | ____ | ☐ | ☐ | ☐ | ☐ | ___ days | ___ days | ☐ | ☐ | ☐ | ☐ | ☐ | ☐ |
| ____ | ____ | ____ | ___/___ month/day | ____ | ____ | ☐ | ☐ | ☐ | ☐ | ___ days | ___ days | ☐ | ☐ | ☐ | ☐ | ☐ | ☐ |
| ____ | ____ | ____ | ___/___ month/day | ____ | ____ | ☐ | ☐ | ☐ | ☐ | ___ days | ___ days | ☐ | ☐ | ☐ | ☐ | ☐ | ☐ |
| ____ | ____ | ____ | ___/___ month/day | ____ | ____ | ☐ | ☐ | ☐ | ☐ | ___ days | ___ days | ☐ | ☐ | ☐ | ☐ | ☐ | ☐ |
| ____ | ____ | ____ | ___/___ month/day | ____ | ____ | ☐ | ☐ | ☐ | ☐ | ___ days | ___ days | ☐ | ☐ | ☐ | ☐ | ☐ | ☐ |
| ____ | ____ | ____ | ___/___ month/day | ____ | ____ | ☐ | ☐ | ☐ | ☐ | ___ days | ___ days | ☐ | ☐ | ☐ | ☐ | ☐ | ☐ |
| ____ | ____ | ____ | ___/___ month/day | ____ | ____ | ☐ | ☐ | ☐ | ☐ | ___ days | ___ days | ☐ | ☐ | ☐ | ☐ | ☐ | ☐ |
| ____ | ____ | ____ | ___/___ month/day | ____ | ____ | ☐ | ☐ | ☐ | ☐ | ___ days | ___ days | ☐ | ☐ | ☐ | ☐ | ☐ | ☐ |
| ____ | ____ | ____ | ___/___ month/day | ____ | ____ | ☐ | ☐ | ☐ | ☐ | ___ days | ___ days | ☐ | ☐ | ☐ | ☐ | ☐ | ☐ |
| | | | | | **Page totals ➤** | ____ | ____ | ____ | ____ | ____ | ____ | ____ | ____ | ____ | ____ | ____ | ____ |
| | | | | | | | | | | | | Injury (1) | Skin disorder (2) | Respiratory condition (3) | Poisoning (4) | Hearing loss (5) | All other illnesses (6) |

*Be sure to transfer these totals to the Summary page (Form 300A) before you post it.*

Page ___ of ___

Public reporting burden for this collection of information is estimated to average 14 minutes per response, including time to review the instructions, search and gather the data needed, and complete and review the collection of information. Persons are not required to respond to the collection of information unless it displays a currently valid OMB control number. If you have any comments about these estimates or any other aspects of this data collection, contact: US Department of Labor, OSHA Office of Statistical Analysis, Room N-3644, 200 Constitution Avenue, NW, Washington, DC 20210. Do not send the completed forms to this office.

OSHA's Form 300A (Rev. 01/2004)

# Summary of Work-Related Injuries and Illnesses

Year 20__ __

U.S. Department of Labor
Occupational Safety and Health Administration

Form approved OMB no. 1218-0176

*All establishments covered by Part 1904 must complete this Summary page, even if no work-related injuries or illnesses occurred during the year. Remember to review the Log to verify that the entries are complete and accurate before completing this summary.*

*Using the Log, count the individual entries you made for each category. Then write the totals below, making sure you've added the entries from every page of the Log. If you had no cases, write "0."*

*Employees, former employees, and their representatives have the right to review the OSHA Form 300 in its entirety. They also have limited access to the OSHA Form 301 or its equivalent. See 29 CFR Part 1904.35, in OSHA's recordkeeping rule, for further details on the access provisions for these forms.*

## Number of Cases

| Total number of deaths | Total number of cases with days away from work | Total number of cases with job transfer or restriction | Total number of other recordable cases |
|---|---|---|---|
| _____ (G) | _____ (H) | _____ (I) | _____ (J) |

## Number of Days

| Total number of days away from work | Total number of days of job transfer or restriction |
|---|---|
| _____ (K) | _____ (L) |

## Injury and Illness Types

Total number of . . .
(M)

| | | | |
|---|---|---|---|
| (1) Injuries | ____ | (4) Poisonings | ____ |
| | | (5) Hearing loss | ____ |
| (2) Skin disorders | ____ | (6) All other illnesses | ____ |
| (3) Respiratory conditions | ____ | | |

**Post this Summary page from February 1 to April 30 of the year following the year covered by the form.**

Public reporting burden for this collection of information is estimated to average 58 minutes per response, including time to review the instructions, search and gather the data needed, and complete and review the collection of information. Persons are not required to respond to the collection of information unless it displays a currently valid OMB control number. If you have any comments about these estimates or any other aspects of this data collection, contact: US Department of Labor, OSHA Office of Statistical Analysis, Room N-3644, 200 Constitution Avenue, NW, Washington, DC 20210. Do not send the completed forms to this office.

### Establishment information

**Your establishment name** ________________

Street ________________

City __________ State ____ ZIP _____

Industry description (*e.g., Manufacture of motor truck trailers*)

________________

Standard Industrial Classification (SIC), if known (*e.g., 3715*)

__ __ __ __

OR

North American Industrial Classification (NAICS), if known (e.g., 336212)

__ __ __ __ __ __

### Employment information *(If you don't have these figures, see the Worksheet on the back of this page to estimate.)*

Annual average number of employees ________

Total hours worked by all employees last year ________

### Sign here

**Knowingly falsifying this document may result in a fine.**

I certify that I have examined this document and that to the best of my knowledge the entries are true, accurate, and complete.

________________ ________
Company executive Title

( ) - ________ / /
Phone Date

OSHA's Form 301

# Injury and Illness Incident Report

**Attention:** This form contains information relating to employee health and must be used in a manner that protects the confidentiality of employees to the extent possible while the information is being used for occupational safety and health purposes.

**U.S. Department of Labor**
Occupational Safety and Health Administration

Form approved OMB no. 1218-0176

This *Injury and Illness Incident Report* is one of the first forms you must fill out when a recordable work-related injury or illness has occurred. Together with the *Log of Work-Related Injuries and Illnesses* and the accompanying *Summary*, these forms help the employer and OSHA develop a picture of the extent and severity of work-related incidents.

Within 7 calendar days after you receive information that a recordable work-related injury or illness has occurred, you must fill out this form or an equivalent. Some state workers' compensation, insurance, or other reports may be acceptable substitutes. To be considered an equivalent form, any substitute must contain all the information asked for on this form.

According to Public Law 91-596 and 29 CFR 1904, OSHA's recordkeeping rule, you must keep this form on file for 5 years following the year to which it pertains.

If you need additional copies of this form, you may photocopy and use as many as you need.

Completed by ______________

Title ______________

Phone (____)____-______ Date ___/___/___

***Information about the employee***

1) Full name ______________

2) Street ______________

City ______________ State ______ ZIP ______

3) Date of birth ___/___/___

4) Date hired ___/___/___

5) ☐ Male
☐ Female

***Information about the physician or other health care professional***

6) Name of physician or other health care professional ______________

7) If treatment was given away from the worksite, where was it given?

Facility ______________

Street ______________

City ______________ State ______ ZIP ______

8) Was employee treated in an emergency room?
☐ Yes
☐ No

9) Was employee hospitalized overnight as an in-patient?
☐ Yes
☐ No

***Information about the case***

10) Case number from the *Log* ______________ *(Transfer the case number from the Log after you record the case.)*

11) Date of injury or illness ___/___/___

12) Time employee began work ______________ AM / PM

13) Time of event ______________ AM / PM ☐ Check if time cannot be determined

14) ***What was the employee doing just before the incident occurred?*** Describe the activity, as well as the tools, equipment, or material the employee was using. Be specific. *Examples:* "climbing a ladder while carrying roofing materials"; "spraying chlorine from hand sprayer"; "daily computer key-entry."

15) ***What happened?*** Tell us how the injury occurred. *Examples:* "When ladder slipped on wet floor, worker fell 20 feet"; "Worker was sprayed with chlorine when gasket broke during replacement"; "Worker developed soreness in wrist over time."

16) ***What was the injury or illness?*** Tell us the part of the body that was affected and how it was affected; be more specific than "hurt," "pain," or sore." *Examples:* "strained back"; "chemical burn, hand"; "carpal tunnel syndrome."

17) ***What object or substance directly harmed the employee?*** *Examples:* "concrete floor"; "chlorine"; "radial arm saw." *If this question does not apply to the incident, leave it blank.*

18) ***If the employee died, when did death occur?*** Date of death ___/___/___

Public reporting burden for this collection of information is estimated to average 22 minutes per response, including time for reviewing instructions, searching existing data sources, gathering and maintaining the data needed, and completing and reviewing the collection of information. Persons are not required to respond to the collection of information unless it displays a current valid OMB control number. If you have any comments about this estimate or any other aspects of this data collection, including suggestions for reducing this burden, contact: US Department of Labor, OSHA Office of Statistical Analysis, Room N-3644, 200 Constitution Avenue, NW, Washington, DC 20210. Do not send the completed forms to this office.